TOOLS AND TECHNIQUES IN BIOLOGY

Tools and Techniques in Biology

UMAIYAL MUNUSAMY

Delve
PUBLISHING
www.delvepublishing.com

Tools and Techniques in Biology

UMAIYAL MUNUSAMY

Delve Publishing

2010 Winston Park Drive,

2nd Floor

Oakville, ON L6H 5R7

Canada

www.delvepublishing.com

Tel: 001-289-291-7705
001-905-616-2116

Fax: 001-289-291-7601

Email: orders@arclereducation.com

ISBN: 978-1-77361-058-0 (Hardcover)

About the Author

UMAIYAL MUNUSAMY

Dr Umaiyal Munusamy is a Plant Scientist and currently working as a postdoctoral at Institute of Plantation Studies, University Putra Malaysia (UPM), Malaysia. She is also an affiliate member of the Young Scientist Network (YSN) Academy Science Malaysia (ASM). Her interest is mainly in science communication, plant biotechnology, transgenic research and plant microbe interactions. Currently she is also keen into looking at sustainable agriculture processes and technologies in accordance with the climate change. Besides research, her passion in teaching continues into educating refugee children voluntarily. Committed to science communication Dr Umaiyal Munusamy Engaged in Kuala Lumpur Engineering Science Festival (KLESF) 2016, had delivered talk on Introduction of Agriculture Engineering to Secondary School Student, performed Science Show at FameLab 2017 competition, coached contestant who participating in scientific communication competition. Last but not least, her passion in writing continues through this book.

Contents

Preface

We regularly refer to items used in gardens as gardening tools, those used in a kitchen as utensils and those used in laboratories as apparatus. This is because tools that are used in a specific field or activity have different designations. A tool is defined as a device or implement, especially one held in the hand, used to carry out a function. We can usually classify a tool based on its function such as cutting and edge tools, moving tools, tools that enact chemical changes, perception tools, shaping tools, fastening tools, information and data management tools and simple machine tools. In addition, we also use to call certain tools as instrument, implement, machine and devices.

What is equipment? Equipment is a set of tools used to achieve a goal. Therefore, these are the best answered differentiation between tools and equipment. According to Oxford dictionary, equipment is defined as solid things that are needed for a particular purpose or activity. Equipment can be presented as a set of tools, a number of tools or a collection of tools. A slight difference is that equipment also contains guards that we use for our safety.

In this book, there will be 7 main topics to be discussed:

1) Tools and Techniques-Laboratory
2) Tools and Techniques-Sampling Biological Sample
3) Tools and Techniques-Storage Biological Sample
4) Tools and Techniques-Processing Biological Sample
5) Tools and Techniques-Biological Sample Analysis
6) Tools and Techniques-Interpreting Biological Data
7) Case Studies

Author

CHAPTER

1 BASIC TOOLS IN LABORATORY

INTRODUCTION

Laboratories are spaces that are used for scientific research. There are many types of laboratories that are differentiated based on the purpose of the laboratory's usage. There are wet laboratories such as a physics laboratory, chemical laboratory, molecular biology laboratory, animal laboratory, medical laboratory and also dry laboratories such as a computer laboratory, biocontainment laboratory and also support system laboratory.

Figure: An example of scientific laboratory

Scientific laboratories can be found as a research and learning spaces in schools, universities, industry, government, military facilities and even aboard ships and spacecraft. In all this laboratories there are different type of tools and equipment are needed. Before discussing more detail on the tools in laboratory, lets discuss the basic tools that are essential to be present in the laboratory in order for the whole laboratory to function efficiently.

A complete tool box is essential to be kept in each laboratory. This tool box will be part of other sophisticate tool to achieve certain purposes. It can also use to repair certain instrument in the laboratory in another word tools helping tools (instruments).

BASIC TOOLS

Screwdrivers

Each laboratory needs to have a screwdriver. If the laboratory storage is limited, buy a screwdriver with interchangeable tips. Otherwise, choose a set of separate screwdrivers in sizes you'll use frequently. If there is enough budget, get a full range of sets. Usually three sizes of slotted-screw drivers are advisable.

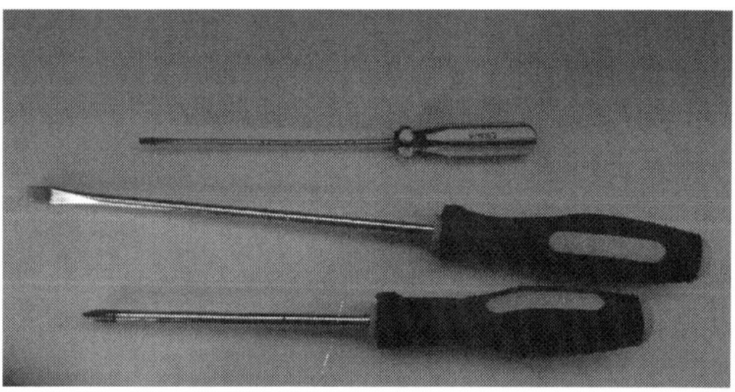

Hammer

A 16-ounce claw-style hammer will handle most jobs. It also comes with a cushioned grip.

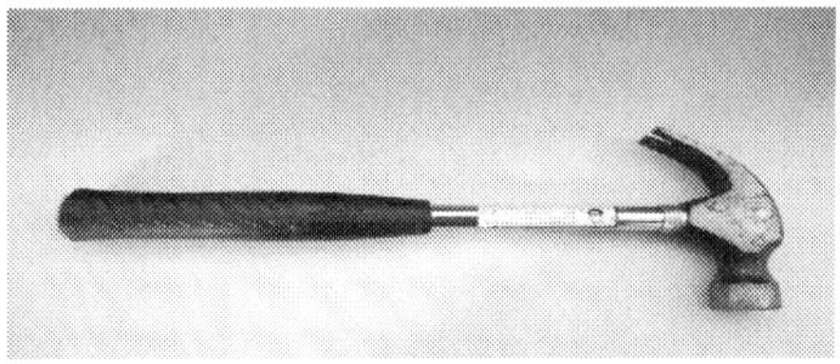

Measuring Tape

Get a 25 foot model with a blade lock. The blade should be at least 1 inch wide to keep it from collapsing when extended beyond a few feet.

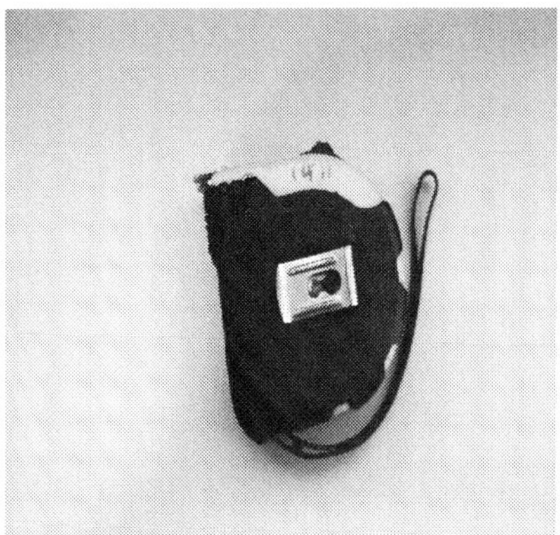

Utility Knife and Blades

A basic model with a retracting blade will be suitable. You may get a non-slip handles and tool-free blade changes and a supply of replacement blades to avoids using a dull blade. Researchers can packed this together while heading to field sampling in the forest.

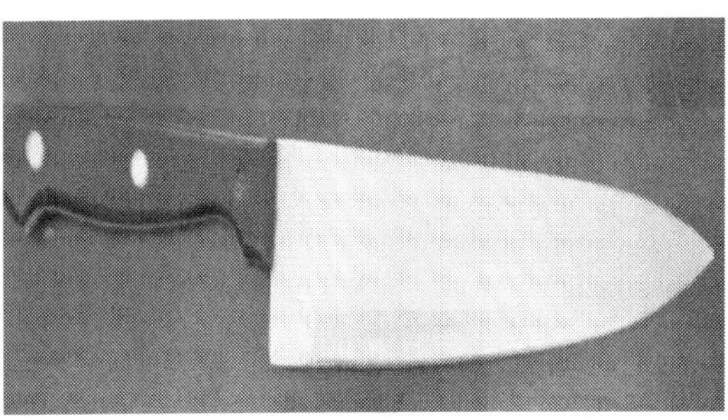

Pliers

The must have models of pliers is 8 inch needle-nose pliers and 10 inch groove-joint pliers. Groove-joint pliers are particularly useful for plumbing projects because of their wide-opening jaws. Laboratory that have sink may need this. You may also get pliers with padded grips. A pair of locking pliers will allow you to focus more strength on turning jammed screws.

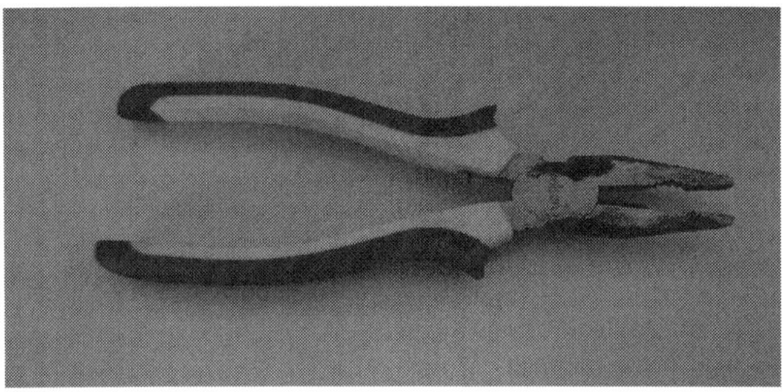

Adjustable Wrenches

These tools will yield jaws less likely to slip and damage your fasteners.

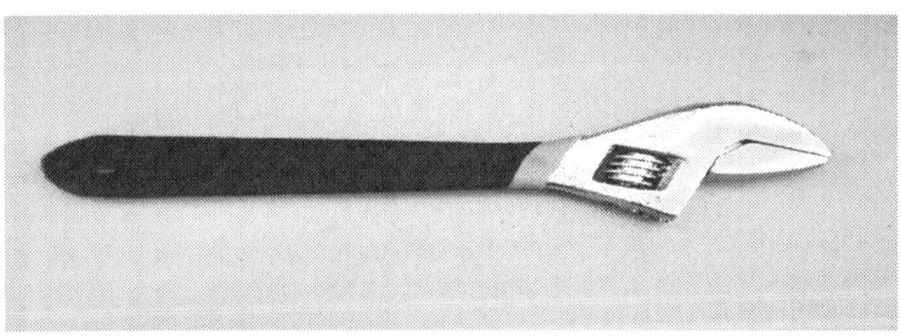

Level

Is use for leveling everything from pictures to a machine. This will help the machine to have a stable ground.

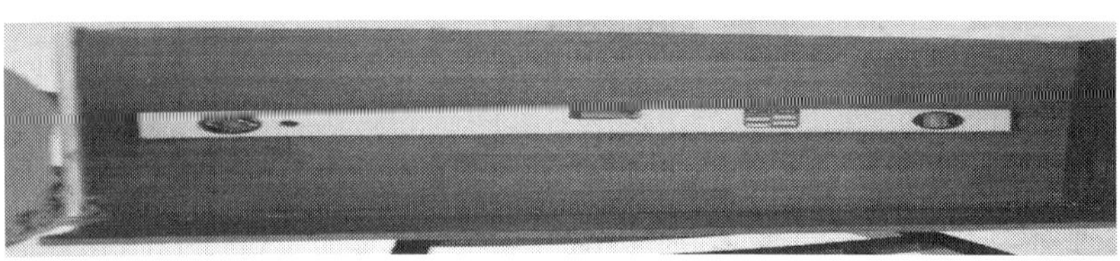

Stiff-Bladed Putty Knife

A 2-inch-wide blade can handle everything for cuttings. Users can packed this while planning for field sampling.

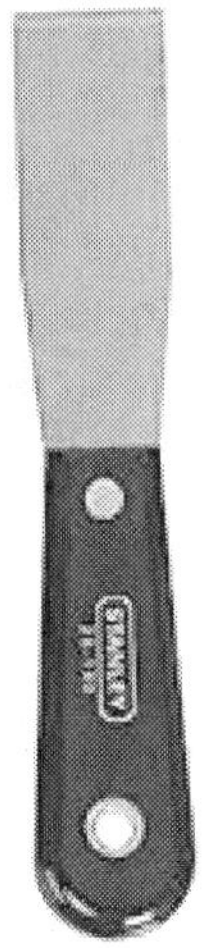

Straight Edge

Start with a 3-foot version and add longer ones as your projects require. The wider the body, the less it will flex during use.

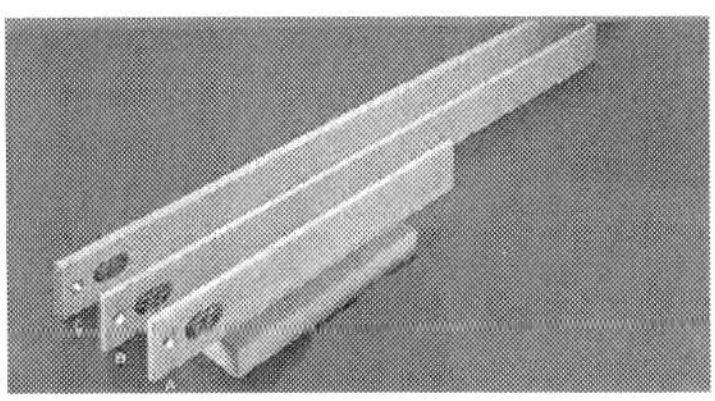

Square

Whether you're setting up tools or marking lines to cut, you'll need an accurate square. Popular types include a try square (also called a "miter square") and the triangular square. The type doesn't matter as much as its accuracy.

Inaccurate squares will lead to inaccurate work and tool adjustments. To check whether your square lives up to its name, hold it against the straight edge of a notepad or a piece of wood with a straight edge. Place the square along the long edge at its center and mark a pencil line dividing the surface in half. Then flip the square over, place the blade as close as possible to the first lineand and draw another line. If the two lines aren't parallel, try another square.

Flashlight/Torchlight

Any sturdy flashlight will do for starters, but you'll soon wish you had one you don't have to hold while you work. Options include flashlights that can be tilted and aimed where you need light and ones with a headband for hands-free aiming. Users can use this while heading to field sampling at the area with less sunlight.

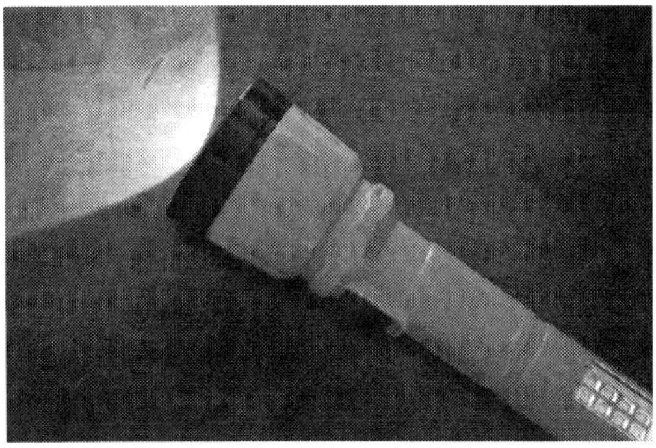

Mill File

Eventually, you'll come across some dangerously sharp metal edges that need to be dulled or dull edges that need to be sharpened (such as a lawn mower blade). A 6-inch mill file offers two flat faces. To keep them in good shape, clean them with a stiff-bristle or brass-bristle brush.

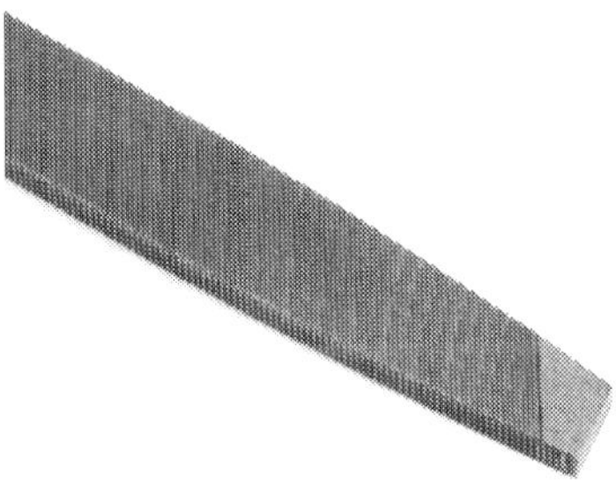

Safety Glasses, Gloves, Hearing Protection And Breathing Protection

Accidents do happen in laboratory. To keep them from happening to you, make a habit of wearing safety equipment. Even if you wear eyeglasses with impact-resistant lenses, flying debris could damage them and even find its way to your eyes. If you don't wear glasses, today's safety glasses can protect you. If you do wear glasses, consider a pair of goggles large enough to slip around them. As you add power tools, you'll want either in-ear or muff-style hearing protection. When working in dusty conditions, a respirator provides a more comfortable way to protect your lungs than paper masks.

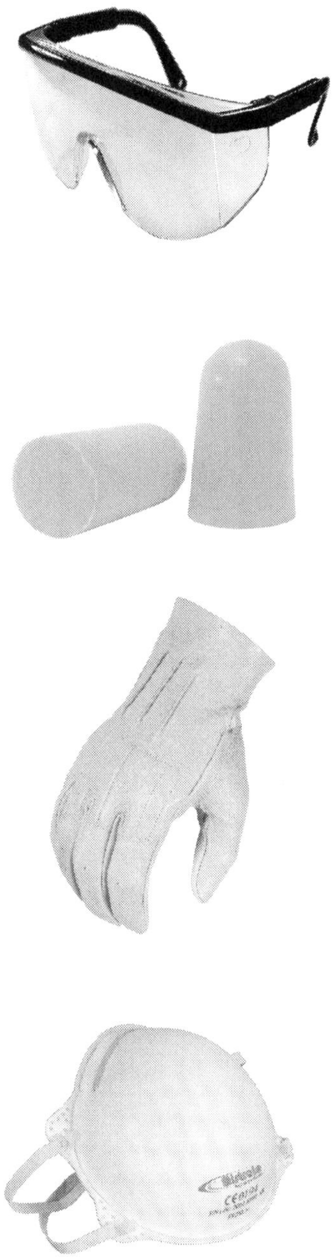

Figure: Safety equipments

Power Drill

You can find both corded and cordless drills to fit any budget. You can choose a cordless model between 12 volts (for small jobs) and 18 volts (for any other job). For laboratory purposes you may pick up a corded drill and a heavy-duty extension cord. You'll get more power at a lower cost and won't have to wait for a battery to charge.

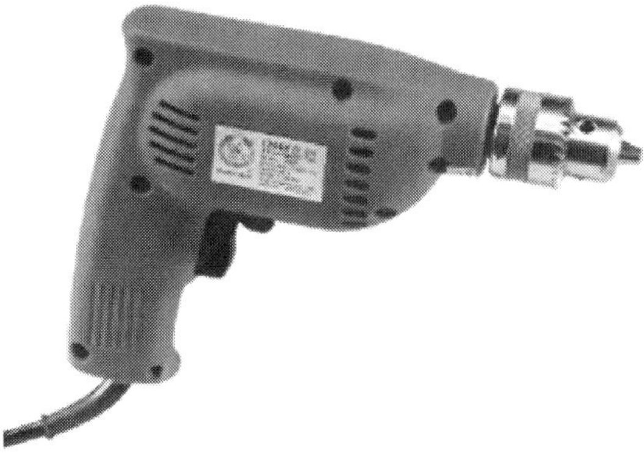

Saw

For the laboratory that involves in sampling at the forest they may need a miter saw or table saw. 7-1/4-inch circular saw is just fine for most purposes. Equipped with blades such as a 24-tooth blade for rough works, a 40-tooth blade works for finer cuts in hardwoods and with less splintering and 180-tooth blade works for smooth cuts of medium density fiber materials.

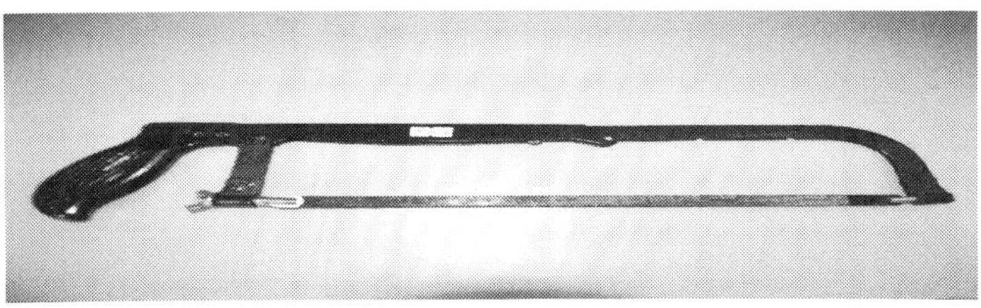

Sawhorses

After a few projects, you'll soon get tired of working on the ground or on a tipsy card table. A simple pair of sawhorses can provide a steady surface at a comfortable level.

Combination Wrenches

Wrenches sets come in two types: SAE (or standard) wrenches are sized in inches (1/4 inch and 1/2 inch and metric wrenches sized in millimeters). Choose SAE wrenches if you'll be working on fasteners around the laboratory. For equipment repairs, though, you may need a metric set or both. The reason they are called combination wrenches because one end is enclosed and studded with "points" that grip the corners of a bolt or nut. The other end is open and useful when you can't slip the closed end in place.

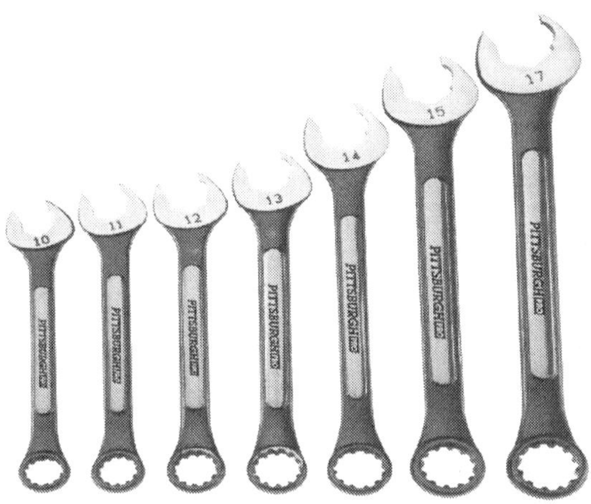

Clamps

Each laboratory can have ones sized of clamps for the types of repairs you'll do the most. A couple of 24 inch bar clamps that you can tighten with one hand will handle many medium-size projects. If most of your repairs are minor, pick up some with smaller versions.

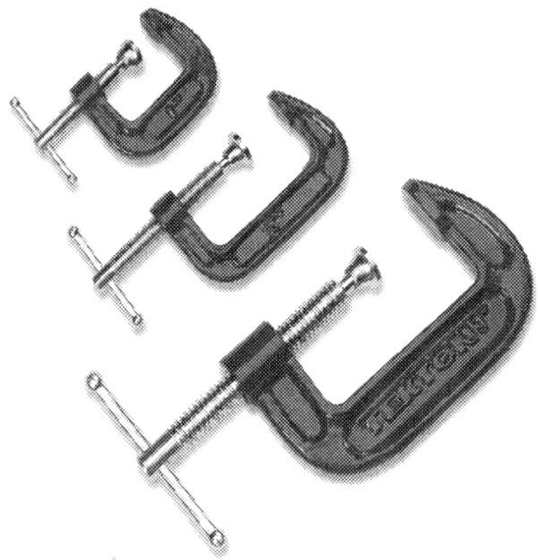

Sander

The best sander for most projects is a random-orbit sander with a 5-inch pad. For most work, you'll need sanding pads ranging from 80-grit to 220-grit sandpaper.

Shop Vacuum

A laboratory can have a vacuum to do cleaning on dust removal or waste materials derived from sampling.

Wire Cutters

Many types of pliers have these built in, but they're seldom as handy to use or durable as a set of sole wire cutters.

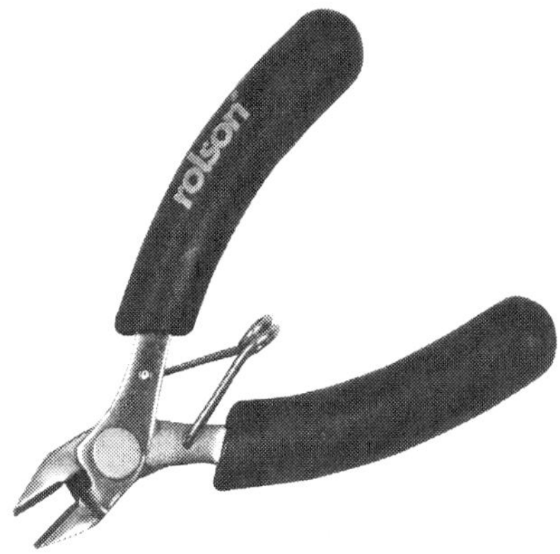

Plunger

All laboratory should have a plunger to clear blockages in sink. It consists of a rubber suction cup with an attached stick made of wood or plastic.

Remove the pop-up drain stopper, if your drain has one. Fill the clogged sink with enough water to completely cover the rubber part of the plunger. Some drain stoppers can be removed by unscrewing them from the top.

Others need to be disconnected from the pivot rod, which is located under the sink. Pull the pivot rod back to disconnect it from the drain stopper. Block any extra openings. This will ensure that you are able to create enough pressure to remove the clog while plunging.

Stuff a wet rag part way into the overflow opening. Enlist the help of another person to hold a wet rag over any opening. Place the plunger directly over the drain to create a seal.

Plunge the sink. Grip the handle of the plunger firmly with both hands, one above the other. Press down as forcefully as possible, then quickly let the plunger come back up. Repeat the up and down motion as rapidly and forcefully as possible.

Remove the plunger by quickly breaking the seal. Check to see whether the clogged water drains. If not, repeat the plunging process until the clog is removed.

Sealing Tape

Sealing tape is also known as Thread seal tape (PTFE tape). It is a polytetrafluoroethylene (PTFE) film for use in sealing pipe threads. The tape is sold cut to specific widths making it easy to wind around pipe threads. Thread seal tape lubricates allowing for a deeper seating of the threads and it helps prevent the threads from seizing when being unscrewed. The tape also works as a deformable filler and thread lubricant, helping to seal the joint without hardening or making it more difficult to tighten and instead making it easier to tighten. Thread sealing tape is almost entirely applied by hand.

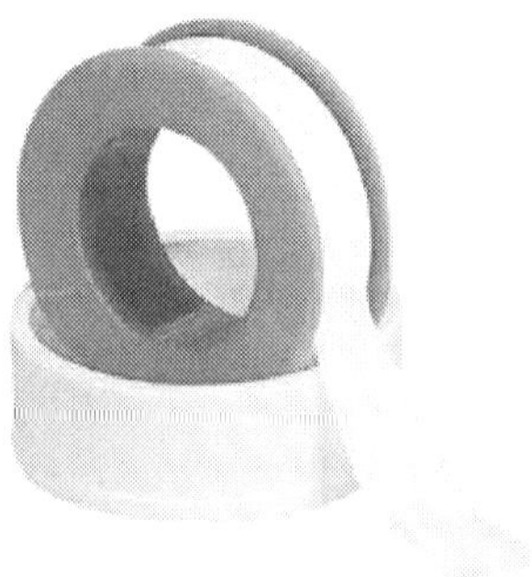

CHAPTER

2 TOOLS AND TECHNIQUES -SAMPLING BIOLOGICAL SAMPLE

INTRODUCTION

Biology is a natural sciences that studies the life of living organisms. Living organism is broadly specified as plant and animal kingdom. In order to study this, the first attempt that we need to do is to go for sampling. Sampling is done because researcher cannot gather data from the entire population. Even in relatively small populations, the data may be needed urgently and including every single unit in the population in the data collection may take too long. In this section, we will discuss the necessary tools and techniques needed during sampling living organism. The purpose of sampling is to determine the population within an area of experiment. In order to obtain a perfect sampling, a proper sampling tool and techniques need to be applied based on the geographical location and the type of samples that being sampled.

The easier way to prepare a perfect sampling tools and techniques is by asking yourself this 3 main questions:

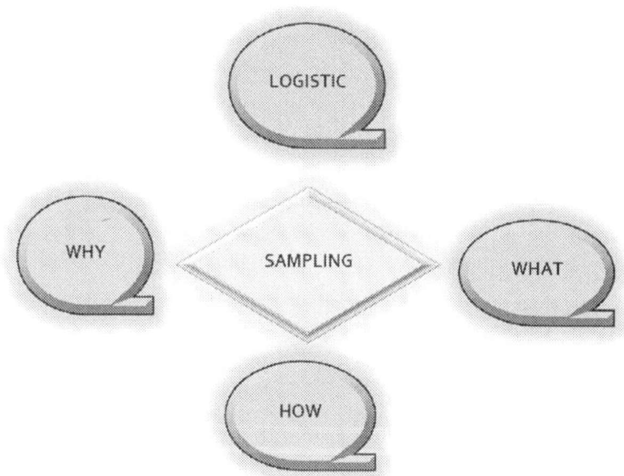

The answer for this question will allow you to select the best tools to use to accomplished researcher`s objective.

Some Examples:

For nutritional sampling, it is helpful to sample from healthy plants as well as affected specimens. Through this you can correlate visible symptoms with lab results and confirm or eliminate suspected causes. In addition, you also should have your soil to be analyzed. This allows you to determine whether a nutritional problem is due to soil deficiency or poor plant uptake.

BASIC SAMPLING TOOLS

There are several basic tools that need to be packed by all sampler prior sampling to any location. These basic tools will help you to find coordinate, for your own safety and to ease your sampling throughout the sampling task.

Compass

It is an instrument used for navigation and orientation that shows direction relative to the geographic. The compass rose shows the directions to the north, south, east and west on the compass face as abbreviated initials. When the compass is used, the rose can be aligned with the corresponding geographic directions. For example, the "N" mark on the rose really points to the north. North corresponds to zero degrees and the angles increase clockwise, so east will fall at 90 degrees, south will be at 180 and west is at 270.

In these days, you can use compass application that is available in your smart phone. But before you can use the apps effectively, you must calibrate your phone's magnetic sensor. To calibrate your phone, hold it up and move it around in a figure 8 fashion. While calibrating, make sure you are far from any electronic devices such as computers, electric fans, Wi-Fi routers, or other electronics that have the possibilities to interfere with the magnetic sensor and its readings in your phone. In addition, material such as metal, rings or jewelry that are close to your phone are essential to avoid.

GPS

GPS also known as Global Positioning System navigation. It is used to record the exact location of a sample. This helps in locating the same group of samples again.

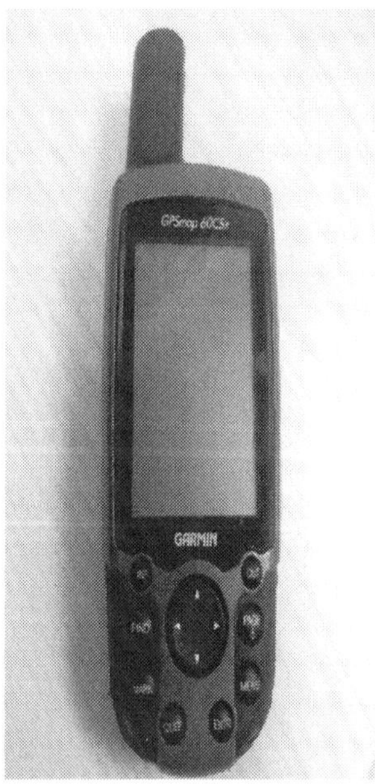

Stationary

Many of us might forget to pack stationary while heading to field sampling. A complete set of stationaries such as pencils, marker pen, stapler, staples, stick note, labels are extremely essential and need to be carry while sampling.

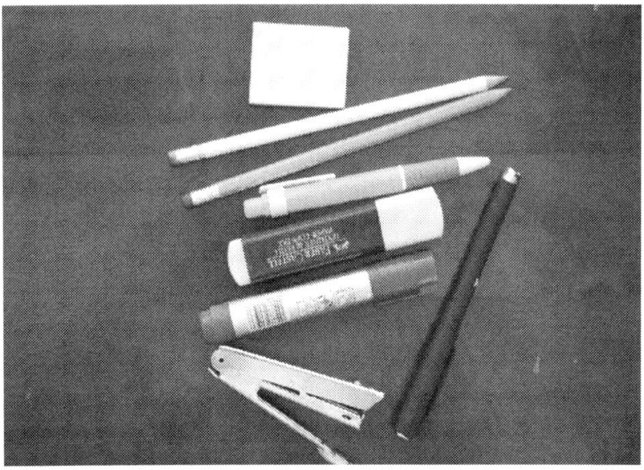

Recording Aids

Recording tools will be similar to any types of sampling. It is very necessary to keep a good track record while sampling so that loss of data was not encounter. Recording aid can be movie cameras, videotape cameras, monitoring systems, tape recorders, microphone, locomotor activity monitoring systems, sonographs and event recorders that aid as behavioral testing devices. These equipment, will help in capturing the behavior of moving samples for example animals in their natural habitat. If you want to make a photograph of the sample collection, you will want to use a camera that is at least 2 to 3 megapixels and has an optical zoom of at least x10 (for the small insects). If the sample is only a few millimeters in size, an optical zoom of x17 will be required. Have a background either white or black, to get a good contrast photographing.

Barcode

A barcode is series of bars and spaces arranged according to a set of rules that determines how data is to be represented. Different bars and spaces patterns are used to express different symbols. These symbols are readable only by a scanner. Barcode is an important identification tool that provides an accurate and timely support of the data requirement for proper management systems. It need to be carried during the sampling task because as you sample, the barcode is labelled on the sample immediately. Barcode technology is an important identification tool that provides an accurate and timely support the data requirement for proper management systems.

Hand Clippers/Knife

Use hand clippers to collect hardy plant samples.

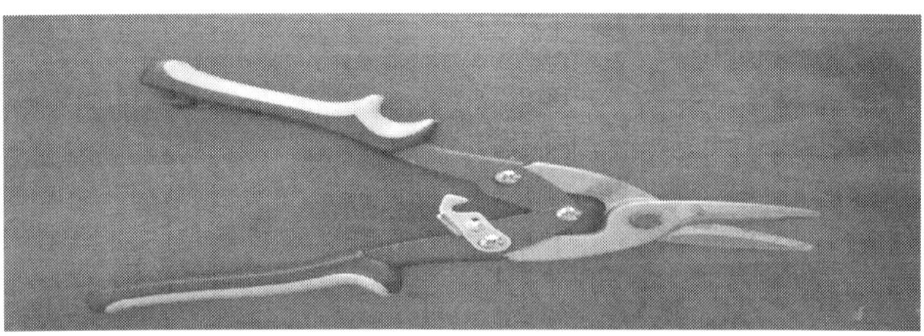

Bags

Ensure that dry samples are air dry (free of surface moisture). Then place them in a paper bags. Do not use plastic bags, which quickly can cause samples to become wet and rot. Wet sample probably good to be kept in the vial (closed container).

Computer

Computers help scientist in maintaining databases of information about their research. The notes that they make when they are in the field are also recorded in these databases for the other researcher to refer and make inferences. Computer tools are also needed for those who use geographic information systems to find or mark a location with or without GIS software, as you may need to create maps using a software such as ESRI ArcView. You can create computer models to demonstrate changes of before and after of a population or an environment that was studied, so you need to use computer programs to generate models or statistical reports.

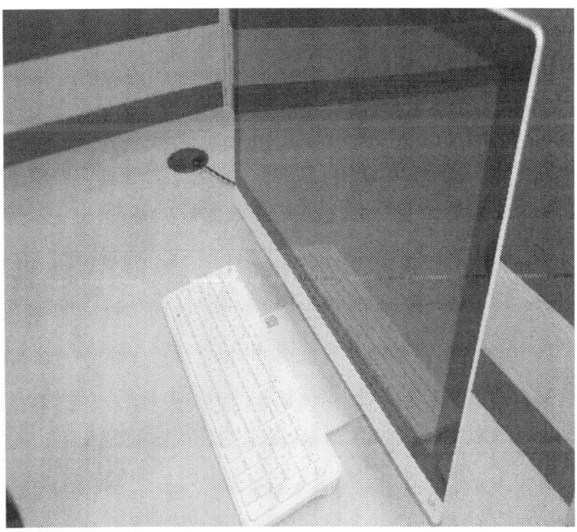

Newspaper or Paper Towels Or Aluminum Foil

For wrapping and ensure that dry sample is air dried.

Hoe

Is needed if you are sampling plant root deep in the soil. You will need hoe to dig the soil out and to dig up the plant root.

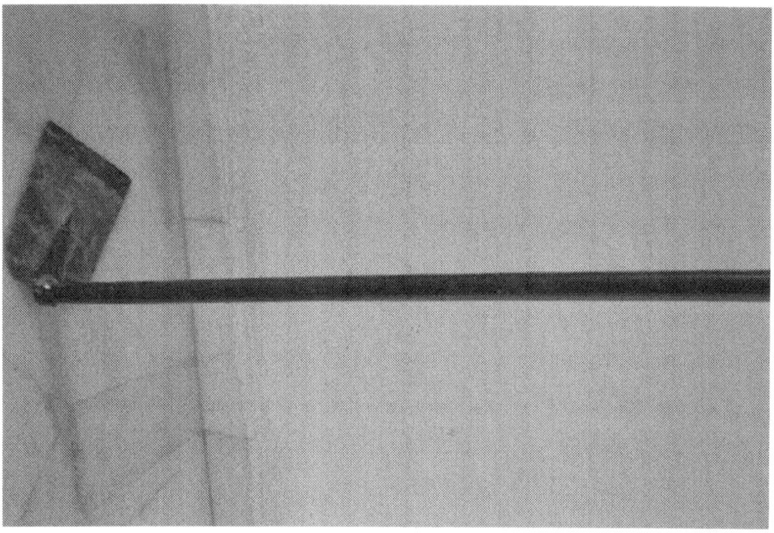

Tents

If you are travelling to remote locations to conduct field studies, or if you need to stay overnights to do samplings, you may need tents too.

Logistics

Proper shipping is an important aspect of submitting samples. Never pack samples in plastic unless you can deliver the sample by hand or with an overnight service and also confirm that a technician will open the sample immediately after it arrives at the lab. If not, use paper bags, paper towels or newspaper. Use sturdy packaging that will protect the sample from damage during shipping. Overnight shipping is preferred, but if you do use regular mail, send the sample early in the week. Late-week mailings run the risk of sitting in a warm post-office over the weekend.

TOOLS AND TECHNIQUES IN PLANT SAMPLING

Introduction

There many types of plants exist in the ecosystem. Botanist classified plant into groups that have similar characteristics. The basic reasons for sampling plant is to obtain a nutritional analysis, to diagnose a disease problem and to determined population of the said sample or to identified and characterized novel sample in the ecosystem. There are 7 major plant groups that can be sampled such as:

Bryophytes	Liverworts/Green Algae

For disease-diagnosis samples, do not apply pesticides before collecting samples, because this could kill the organisms, such as fungi, that you are trying to identify. Collect samples from areas in several stages of disease development and especially make sure you include samples from transition areas between healthy and disease sample. Keep the soil-and-root layer intact. Sample the largest sections of the plant. Shoot sections from woody plants should be at least 6 to 8 inches long. Sample as many portions of the plant as possible, including flowers, fruit, leaves and roots.

Ecological Sampling

The common sampling tools that is used for sampling ecological environment is known as quadrats. Quadrats normally consist of a square frame sized 1m^2, rectangular and circular. The purpose of using quadrats is to enable the researchers to compare samples obtained from areas of consistent size and shape. It is common that data derived from square quadrat are only comparable from data derived similar square quadrats.

The choice of quadrat size depends on the type of experiments. Small quadrats are much quicker to analysed but are less reliable than the large ones. While, larger quadrats require more time and effort to examine properly. A balance is therefore necessary between what is ideal and what is practical based on the size of the area that planned to be studied. As a general guideline, any researcher may use

a) 0.5 - 1.0 m^2 quadrats for short grassland or dwarf plant
b) 2 m quadrats for taller grasslands and shrubby habitats
c) 20 m^2 or larger quadrats for forest habitats
d) 0.25 m^2 quadrat for moss species

The sample estimation can be further improved by dividing the quadrat into a grid of 100 squares each representing 1% cover. This can be done by dividing the quadrat by using string or wire attached to the frame in between the existing frame. This is only practical if the vegetation in the area to be sampled is very short, otherwise the string/wire will impede the laying down of the quadrat over the vegetation.

Random Sampling

Random sampling usually was carried out when the area under study is uniform, large and there is limited time available. In this technique, large numbers of samples/records are taken from different positions within the studied habitat. A quadrat frame is most often used for this type of sampling. The frame is placed on the ground and the plants inside it is counted, measured, or collected, depending on the objective of the survey. This is done in triplicate or as many time as possible at different points within the habitat. In the simplest form of random sampling, the quadrat is thrown to fall at 'random' within the site. However, since true randomness is an important element in ecology this is usually unsatisfactory because the capability of the researcher to throw the frame bigger than 1 m^2 quadrat is not truly random.

To applied random sampling within habitats such as woodlands or scrub areas, it is also often not possible to physically lay quadrat frames down, because tree trunks and shrubs will get in the way. In this case, an area with the same size as the quadrat need to be measured out instead and the corners need to be marked to indicate the quadrat area to be sampled. A better method of random sampling is to map the area and then to lay a numbered grid over the map. A (computer generated) random number table is then used to select which squares to sample in.

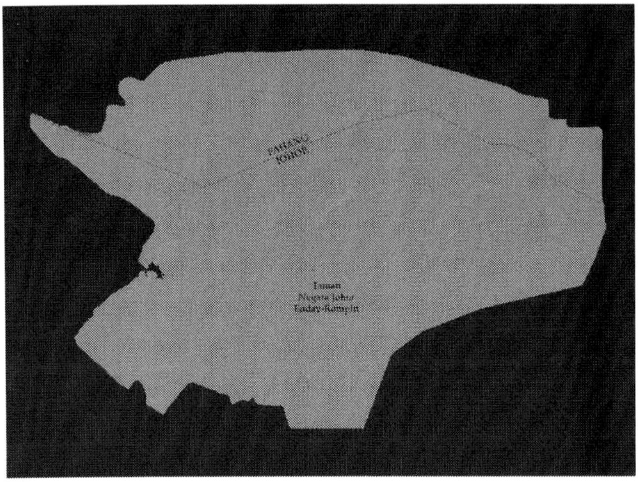

Figure: A grid map of an area to be sampled

Systematic Sampling

Systematic sampling is carried out when samples are taken at fixed intervals along a line. This normally involves doing transects, where a sampling line is set up across an area where there are clear environmental gradients. For example, researchers might use a transect to show the changes of plant species as she/he moved from grassland into woodland, or to investigate the effect on species composition of a pollutant radiating out from a particular source. There are 3 systemic sampling:

a) Line Transect Method

b) Belt Transect Mcthod

c) Stratified Sampling Method

Line Transect Method

A transect line can be made using a nylon rope marked and numbered at each 0.5 m or 1 m intervals in length across the area that was wished to study. The position of the transect line is very important and it depends on the direction of the environmental gradient that is studied. It should be carefully placed otherwise the researcher will end up without clear results. Since in research, time is usually money, so it is important to design the experiment effectively before starting any line transect. A line transect is carried out by unrolling the transect line along the identified gradient. The species touching the line will be recorded along the whole length of the line continuously. At the similar time, the number of species at each marked point is recorded systematically. If there is a slope along the transect line, the results will be also incorporated in the studies as well.

Belt Transect Method

This is similar to the line transect method but gives information on abundance as well as presence, or absence of species. It may be considered as a widening of the line transect to form a continuous belt, or series of quadrats. In this method, the transect line is laid out across the area to be surveyed and a quadrat is placed on the first marked point on the line. The plants and/or animals inside the quadrat are then identified and their abundance estimated. Animals can be counted through this method if they will sit still, or collected, while it is usual to estimate the percentage cover of plant species. Cover is the area of the quadrat occupied by the above-ground parts of a species when viewed from above. The canopies of the plants inside the quadrat will often overlap each other, so the total percentage cover of plants in a single quadrat will frequently add up to more than 100%.

Quadrats are sampled all the way down the transect line, at each marked point on the line, or at some other predetermined interval (or even randomly) if time is short. It is important that the same person should do the estimations of cover in each quadrat, because the estimation is likely to vary from person to person. If different people estimate percentage cover in different quadrats, then an element of personal variation is introduced which will lead to less accurate results. The height of plants in the quadrat can be recorded and the biomass of plants can also be measured by harvesting all the plants inside the quadrat and then weighing either fresh, or dry weight in the laboratory. This is obviously a very destructive method of sampling which could not be used too often in the same place. Sampling should always be as least destructive as possible and you should try not to trample an area too much when carrying out your experiment.

Stratified Sampling

Stratified sampling is used at different areas (or strata) which are identified within the main body of a habitat. These strata are sampled separately from the main part of the habitat.

Individual habitats are rarely uniform throughout their extent. There are often smaller identifiable areas within a habitat which are substantially different from the main part of the habitat. For example, scrub patches within a heathland area, or areas of bracken in a grassland.

One of the problems with random sampling is that random samples may not cover all areas of a habitat equally. To continue with the example of bracken patches in a grassland, if the area was random sampled, it is possible that none of the samples might fall within the bracken patches. The results would then not show any bracken in the habitat. Clearly this would not be an accurate reflection of the habitat. In this sort of situation, stratified random sampling would be used to avoid missing out on important areas of the habitat. This simply means identifying the bracken as a different unit within the habitat and then sampling it separately from the main part of the habitat.

While the bracken area clearly needs to be taken into account, it is nevertheless important to avoid over emphasising its significance within the habitat as a whole. Its importance is to kept in context by locating a proportional number of samples directly within the bracken unit. The proportion of samples taken within the unit is determined by the area of the unit in relation to the overall area of the habitat.

For example, say the grassland area is 200 m^2 overall, with the bracken patch occupying 50 m^2 of this total area. The bracken therefore accounts for 25% of the total grassland area. Say it has been decided that a total of 12 samples need to be taken in order to accurately reflect the composition of the whole habitat. Then 3 of those samples (one quarter, or 25%) would be located within the bracken unit and 9 (three quarters, or 75%) in the general grassland area.

There is a standard formula for calculating the number of samples to be placed in each unit. This is:

$$\text{The area of quadrats sampled in the unit} = \frac{\text{the area of the x the total number of quadrats to be sampled}}{\text{total area of the habitat}}$$

For example, in the illustration given above this would be:

$$3 = \frac{50 \times 12}{200}$$

Tools And Techniques In Agriculture

There are variety of tools that are available for agriculture. Agriculture test kits provides growers with an economical way to quickly test pH as well as the three basic elements needed for a healthier plant which is nitrogen, phosphorus and potassium. The chemical composition of soil includes pH and chemical elements. Soil analysis is necessary for better management of fertilization and to know the residues of fertilizers in relation to the crop, tillage and the most suitable plant choice for soil composition. An analysis can highlight shortages and help the understanding of the causes of an abnormal growth. By using the soil test kits, it is possible to measure pH and the most important element for plant growth (N, P, K). Testing the soil during each crop cycle and comparing the results with plant growth can be useful experiment for subsequent cultivation.

In agricultural application, monitoring the quality of the soil is extremely important for the health and growth of the crop. The pH level is an excellent guide as to which plants may thrive in a particular terrain, as well as indicating which conditioners and fertilizers to use.

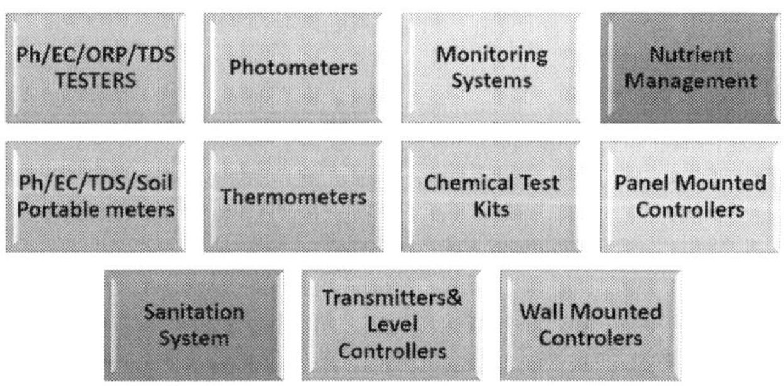

Figure: Some examples of tools used in agriculture

Ec/Tds Testers

EC/ TDS are monitors in drinking water, water conditioning, reverse osmosis, cooling towers, wastewater, laboratories, agriculture, aquaculture and aquariums and hydrophonics. It comes with an amperometric graphite electrode that provides better repeatability in measurements since they do not oxidize. An amperometric graphite measurement of EC/TDS is based on Ohm`s Law, I=V/R where R depends on the distance between two pins and their surface. Oxidation changes both the distance and surface which will affect the accuracy. The non-oxidizing graphite pins are able to provide an optimal surface for accurate and dependable results. When calibration is required simply submerge the electrode tip into calibration solution and adjust the trimmer on the side of the tester.

Temperature Tester

Temperature tester is directly used to measure soil conductivity and temperature with a temperature sensor in the penetration probe. The measurements are automatically temperature compensated.

pH Measurement Kit

It can be used directly in the soil or after preparation of a diluted sample. Use different tips for different type of soil

- rugged tip for moist or soft soil and
- plastic auger for harder ground

Direct Soil Activity And Solution Conductivity Measurement Kit

It is an instrument that has been designed to address the need for fat and accurate conductivity measurements in soils and liquids. It measurers the soil conductivity in EC (mS/cm) as well as salt activity (g/L) directly by inserting the stainless-steel probe. It will illuminate alarm if the tested soil is too dry or nutritive substances such as potassium or nitrogen are lacking. Demineralized water can be added to the soil prior to tests.

Conductivity is an important factor in greenhouses and hydroponics. It is measured in soil as well as in fertilizer solution since it is an excellent indication for the presence of nutritive salts. Soil conductivity is checked before and after fertilization to establish its effectiveness as well as ensuring that the soil is not too saline to damage the plant roots. Conductivity of the irrigation water and fertilizer mixes is checked to make sure values are within an acceptable range and correct fertilizer concentration strength is being applied.

Nutrient Analysis Photometer For Greenhouses And Hydroponics

Deciding which fertilizer to use can be overwhelming, but with a bit of knowledge you can be sure your crop will get the right amount of nutrients it needs. Photometer is a multiparameter bench meter that measures nitrogen, phosphorus and potassium (NPK). Nitrogen is a key factor for fertilization. Nitrogen allows the development of the vegetative activity of the plant, in particular, causes lengthening of trunks and sprouts and increases the production of foliage and fruits. An excess of nitrogen wakens the plants structure creating an unbalanced relationship between leaves and stalks makes the plant become less resistant to disease.

Phosphorus is an important element in the composition of DNA and RNA, the regulators of the energetic exchange (ATP and ADP), as well as the reserve substance in seeds and bulbs. It contributes to the formation of buds, roots, blooming, lignification and size. A lack of phosphorus results in stifling of plants, slow growth, a reduction of production, smaller fruits and a lower expansion of the roots. Potassium helps guard a plant from disease and insects as well as temperature extremes such as drought and cold. It also controls the cellular turgor and the accumulation of carbohydrates. It increases the size of fruits, flavor, color and flagrance of flowers.

Suction Lysimeter For Root Level Soil Monitoring

Lysimeter is an ideal tool for collecting soil solution samples and then perform quantitative chemical analysis. In this way, the operator can easily monitor the level of nutrients, such as ammonia, nitrate, phosphorus and potassium, sulfate, calcium, magnesium. The ceramic tip of lysimeter can be used in all type of soil and it is made of a synthesized material that does not react with nutrient elements. The soil solution therefore, is not affected by the chemical composition of the ceramic cap resulting in precise and reliable tests. Lysimeter allows the extraction of a solution from the soil by creating a vacuum (negative pressure or suction) inside the sampler tube, that exceeds the soil water tensions. This will establish an hydrosolic gradient for the solution to flow through the porous ceramic cap and into the lysimeter tube. Typically, a vacuum of about -60 cb (centibar) should be drawn. For better monitoring of soil solution composition throughout an entire growth period of crops, at least two lysimeters should be installed in the root zone of a representative plant. One at the upper and one in the lower part of the root zone. For better measurement accuracy and repeatability, it is recommended to replicate installing in at least two more location.

TOOLS AND TECHNIQUES IN AQUACULTURE

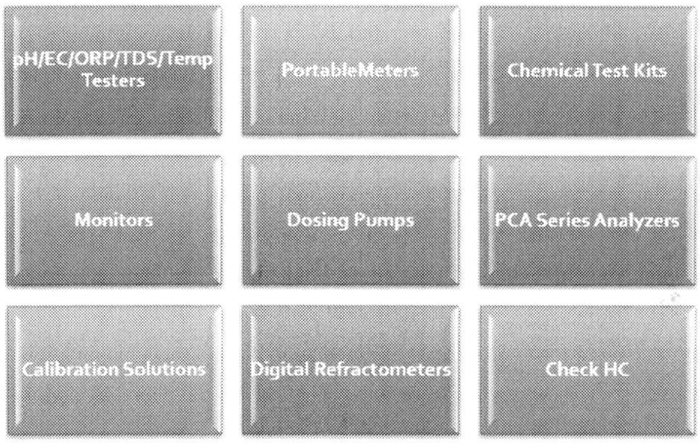

Figure: Some examples of tools used in aquaculture

Chlorine is the most common water disinfectant. The monitoring of chlorine is crucial in applications such as in agriculture, disinfection and water. By monitoring this crucial parameter, serious risks can be avoided.

Usage of HC checkers will actually bridge the gap in simple chemical. Chemical test kits are not very accurate and only give 5 to 10 points of resolution while professional instrumentation can cost hundreds of dollars and can be time consuming to calibrate and maintain. The HC checker used DPD method which approved by EPA and it is easy to use. The HC checker comes in different version within various instrumental companies.

Alkalinity Test Kit

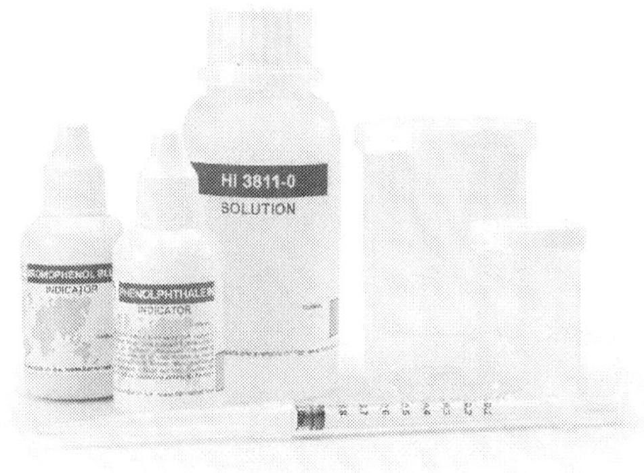

Alkalinity is the quantitative capacity of a water sample to neutralize an acid to a set pH. This measurement is very important in determining the corrosive characteristics of water due to primarily hydroxide, carbonate and bicarbonate ions. Other sources of alkalinity can be from anions that can be hydrolyzed such as phosphates, silicates, borates, fluoride and salts of some organic acids. Alkalinity is critical in the treatment of drinking water, wastewater, boiler & cooling systems and soils.

Alkalinity test kit makes monitoring easy, quick and safe. The compact size gives the user the versatility to use the kit anywhere. The design makes the kit easy to handle and except for alkalinity titrant, practically prevents accidental or damage due to the spills. Alkalinity can be measured as phenolphthalein alkalinity and total alkalinity. The phenolphthalein alkalinity is determined by neutralizing the sample to a pH of 8.3 using dilute hydrochloric acid solution and a phenolphthalein indicator. Since bicarbonate ions can be converted to carbonic acid with additional hydrochloric acid, the phenolphthalein alkalinity measures total hydroxide ions, but only half of the bicarbonate contribution.

Water Purity Testers

Water purity testers enable users to check the purity of distilled or demineralized water in laboratory or industrial environments. It provides fast, on the spot checks for minute traces of contamination in water.

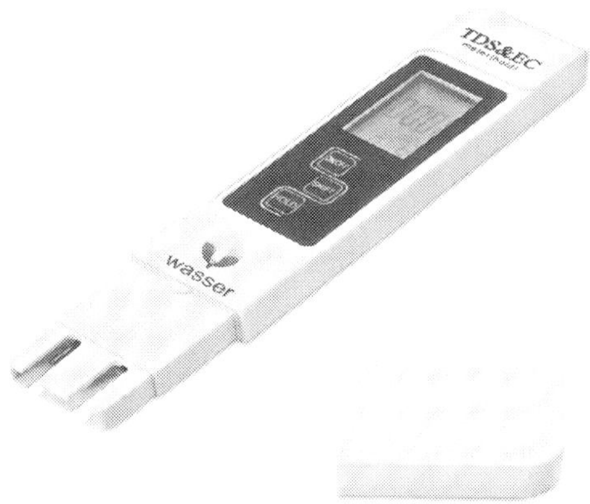

Ammonia Portable Photometers

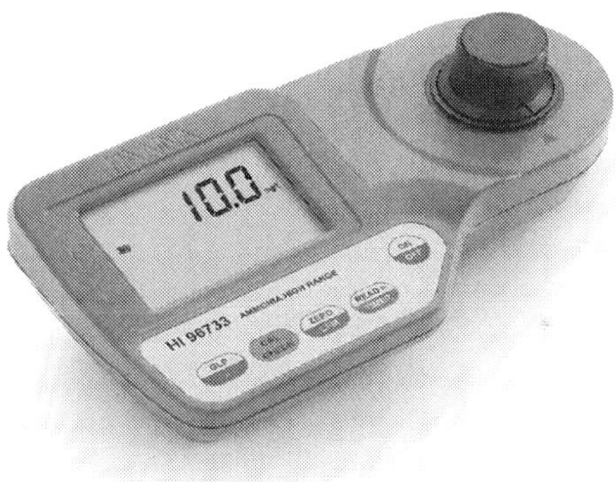

It measures the ammonia-nitrogen content in water sample,waste water and seawater or any type of liquid. These meters use an exclusive positive-locking system to ensure that the cuvette is in the sample place every time it is placed into the measurement cell. Ammonia is often an excellent indication of the presence of animal or plant microbiological decay. It is tested in fish farm (fresh and salt water tanks). Its presence in rivers and reservoirs normally points to agricultural and civil pollutants. Ammonia is tested in lakes, rivers, portable water, boiler feed water, sewage, industrial and waste water. Ammonia-nitrogen in the form of NH_3 and NH_4^+ is often present in water as a component of nitrogen cycle. In the metabolism of proteins and amino acids, many heterotrophic bacteria and fungi (occurring in both soil and water) excrete excess of nitrogen and ammonia. Generally, in unpolluted waters ammonia and ammonium compounds occur in relatively small quantities, on the order of 0.1 mg/L while higher levels usually indicate organic solution. Ammonia is also recognized to be toxic to diatoms in the 7.4-8.5 pH range at a level of 1.1 mg/L and to fish, in the same pH range at a level of 2.5 mg/L.

pH/pH-MV/ORP and TEMPERATURE METERS

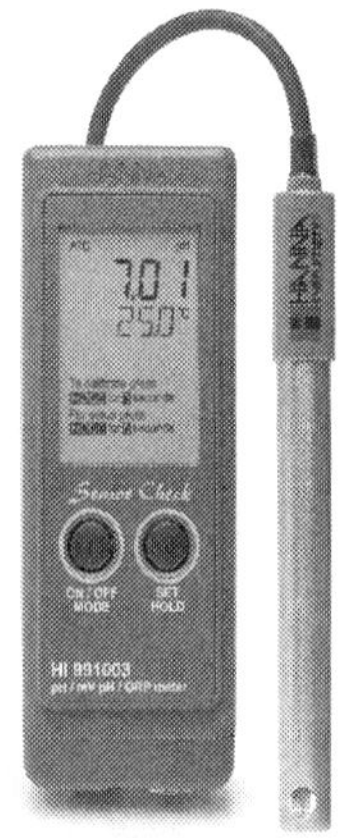

This tool allows users to detect the pH and temperature of any water sample with a high degree of accuracy and fast response. It can fit in the palm of your hand and the probe connection ensures the electrode cable doesn't get in your way. Calibration is automatic at one or two points with three memorized buffer values. Once the instrument has been calibrated, the buffer values used during calibration are displayed with tags on the LCD. This feature keeps users informed of the current calibration and help to avoid taking measurements that are out of range. While the temperature benchtop meter allows users to collect the water sample and to bring back to laboratory for testing. It can measure oxidation reduction potential (ORP) in the medium voltage range. This tools feature manual or automatic temperature compensation with the temperature probe. Good laboratory practice feature provides data consistency. A calibration reminder can be set to alert the user that too much time has elapsed since the last pH calibration and a new one should be performed.

Turbidity Meter

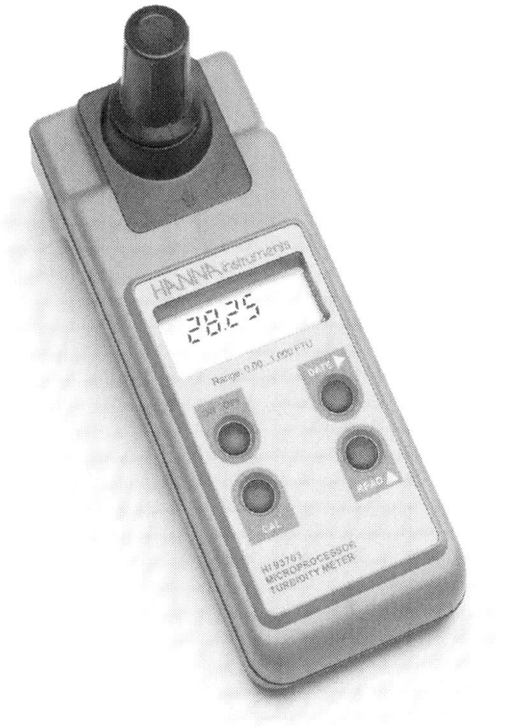

This tool incorporates complete good laboratory practice functions that allow traceability of the calibration conditions. The last calibration time and date can be checked at the touch of a button. The instruments have an EA compliance reading mode which rounds readings to meet EPA reporting requirements. Users will appreciate the accuracy and sensitivity of this instrument, particularly at very low turbidity levels.

Dissolved Oxygen Meter

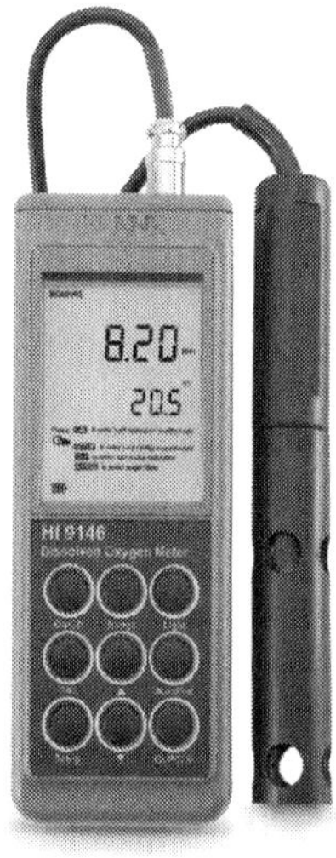

Measures up 45ppm (mg/L) with temperature compensation and automatic calibration. It has been developed for dissolved oxygen and temperature measurement in water, waste water and applications such as fish farming. This tools also allows altitude compensation up to 4000m and the ppm are compensated for changes in the solubility of oxygen in water and for permeability of the membrane as well as the temperature effect.

Digital Refractometer For Salinity Measurement

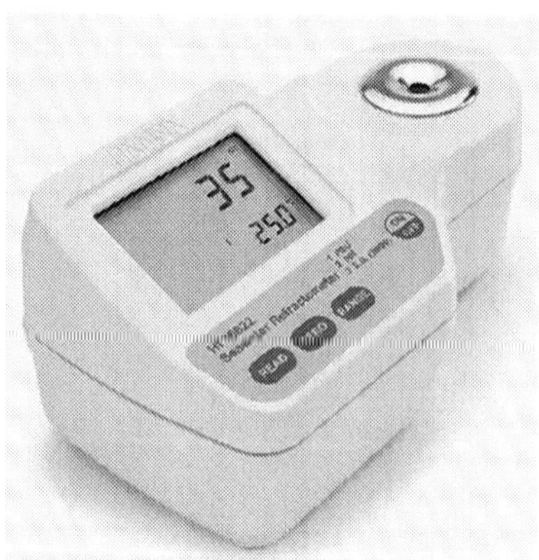

Salinity is a critical measurement in many applications, such as in aquaculture, environmental monitoring, aquariums, desalination plants, well water and many more. Digital refractometers are far better than hydrometers and ocular refractometers in terms of getting accurate reading of the salinity, are less susceptible to changes in temperature, lightweight (can easily dislocate to the field), easy to use, cost efficient, is extremely accurate and its ability to read in all the three of the most widely used salinity units (PSU, ppt and specific gravity). After a simple calibration with distilled or deionized water, a water sample can be introduced into the sample well to be analysed.

Cod Test Tube Heater By Turning Off The Heating Elements

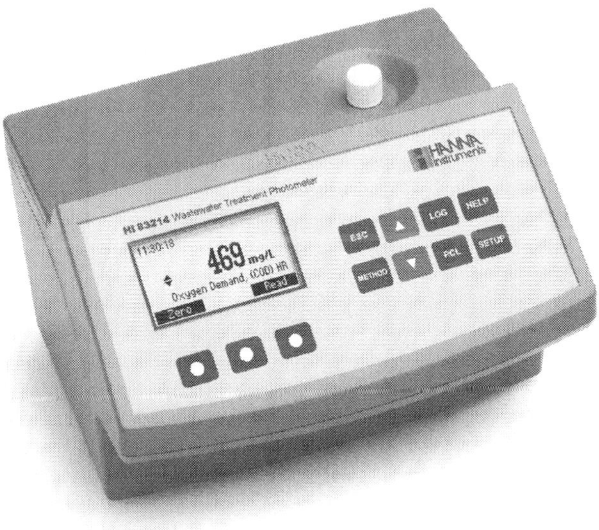

This COD test tube heater is used to heat COD, phosphorus and nitrogen digestions. It contains a 3-hour countdown timer to control timed digestion, a thermal fuse that prevents overheating. Block temperature is continuously displayed on the LCD even when there is no active temperature program running.

Aquatic Net

Courtesy of Sohier (2013)

Aquatic nets can range from student aquatic nets, heavy duty D-style nets to professional triangular aquatic nets as well as to dip nets and fine-mesh nets for collecting small pond organisms such as *Daphnia* to large aquatic organism.

Midwater

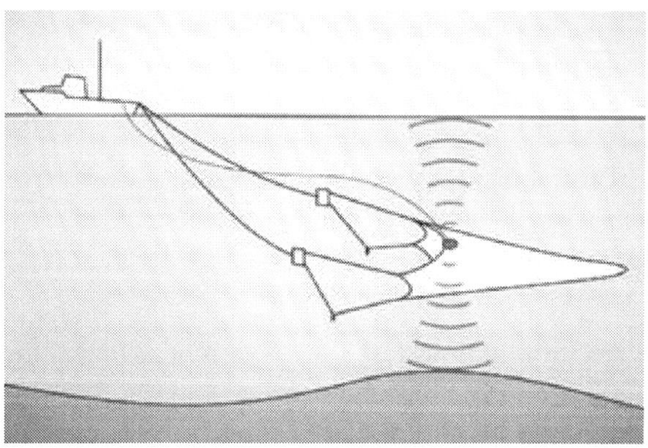

Image courtesy of Food and Agriculture Organization of the United Nations

Researchers use midwater to catch fish that are exist between the sea surface and bottom. It consists of a set of gears. While, researchers use midwater trawls to perform catching in the shallow-surface layer and it are configured with floats. Midwater and midwater trawls have many similarities. However, both differ in dimensions and shapes due to their different fishing targets and hydrodynamic regimes of operation. Unlike midwater, a midwater trawl consists of a cone shaped body which made of four panels with lateral wings extending forward from the opening. Midwater trawls are also designed to catch fish in the midwater column, while maintaining an open net mouth.

Plankton Net

Researchers use plankton nets to collect various size planktonic organisms. It is towed by a vessel. Plankton nets have a long funnel shape and changeable mesh size of the net. At the end of the funnel there is a collection cylinder called a cod-end to collect all the sample.

Ring Net

Image courtesy of sea-gear corporation

Researchers use ring net to collect plankton and other desired particulate. The ring net consists of a fine-meshed bag attached to a metallic ring. The net itself is terminated in a bottle or jar for sample collection. Ring net may be towed for controlling its passage through the water column.

Bongo Net

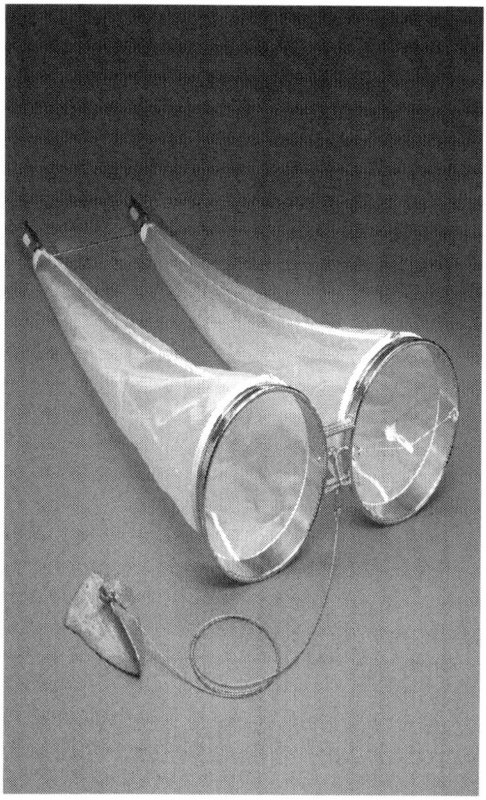

Courtesy of Sohier (2013)

Bogo nets are often used to caught floating fish eggs and newly hatched larvae. The mesh size of a Bogo nets is very fine 20µm up to 1000µm (1mm). This physical characteristic will allow eggs and larvae with sizes from 1-20mm to be trapped perfectly. The user will tow the net from the surface until it reached the bottom of the sea for sampling through out the water column. However, researcher will find difficulties to estimate the quantitative samples of phytoplankton, zooplankton, other invertebrates and large fish as they are unable to measure the volume of water that is filtered during the sampling process. Therefore, to date most bongo and ring nets are deployed with mechanical or electronic flow meters that are positioned in the mouth of the net to quantify the volume of water filtered.

Multiple Opening/Closing Nets And Environmental Sampling System(Mocness)

Courtesy of Sohier (2013)

The Multiple Opening/Closing Nets and Environmental Sampling System, which is identified by its acronym MOCNESS, is widely used for capturing plankton at specific depths on the command of the users. This system carries several sensors for measuring environmental parameters as it is towed. These sensors will measure and record the conductivity, temperature, pressure, fluorescence, optical transmission, dissolved oxygenand light levels.

Neuston Net

Courtesy of Sohier (2013)

Neuston nets are used to collect those organisms associated with the water surface, where they are usually supported by surface tension. While sampling, these types of nets are towed at the surface level of the sea water. Researcher can determine the number of organisms per unit volume of water filtered.

Drift Net

Courtesy of Sohier (2013)

Drift net are used to sample organism such as whales and dolphins. It is type of nets that are not set or fixed but they are 'mobile'. The reason it was named drift is because they are allowed to drift with the prevailing currents occurred in the sea. Therefore, it is used on the high and depth seas for sampled wide range of fish including tuna, squid, salmon and shark.

Gill Net

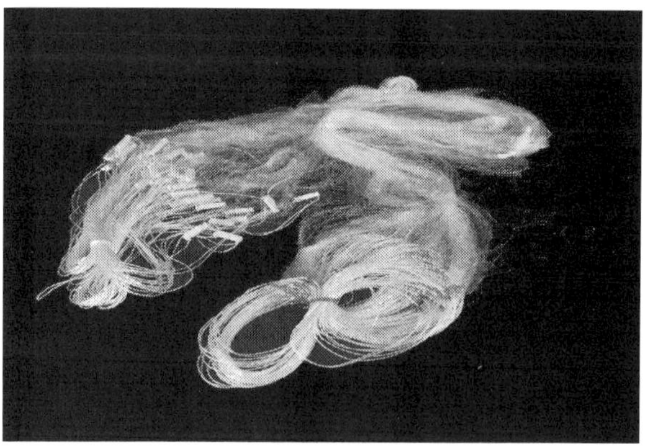

Courtesy of Sohier (2013)

Gill netting is the oldest type of net fishing and is still being used for fish sampling as a traditional tool. Gill nets are suitable to be set below the surface, on the seabed, or at any depth in between. The mesh size used depends upon the species and size range of the fish that being targeted. True gill nets will allow fish to swim through the net, but caught rest of the body. The fish then becomes entangled by the gills as it attempts to back out of the net.

Fyke nets

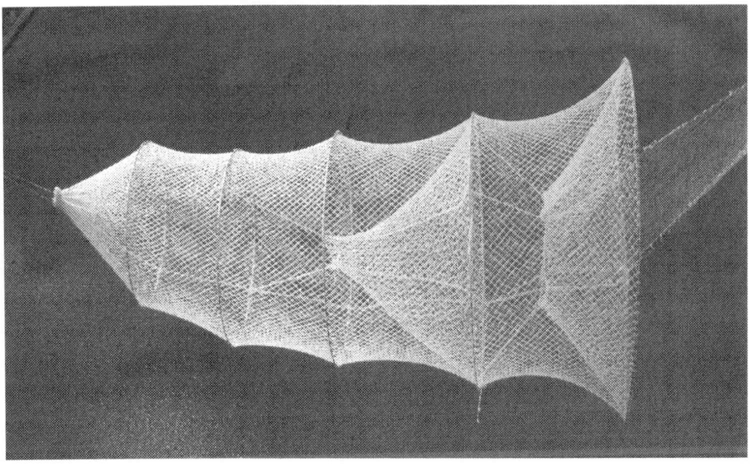

Courtesy of Sohier (2013)

Fyke net or also known as a fish trap is used to sample fish communities that present in a shallow water. It is fixed on the bottom by anchors, ballast or stakes. It consists of cylindrical or cone-shaped netting bags mounted on rings. In addition, it has wings which will guide the fish towards the entrance of the bags and trapped within it.

Hyperbenthos net

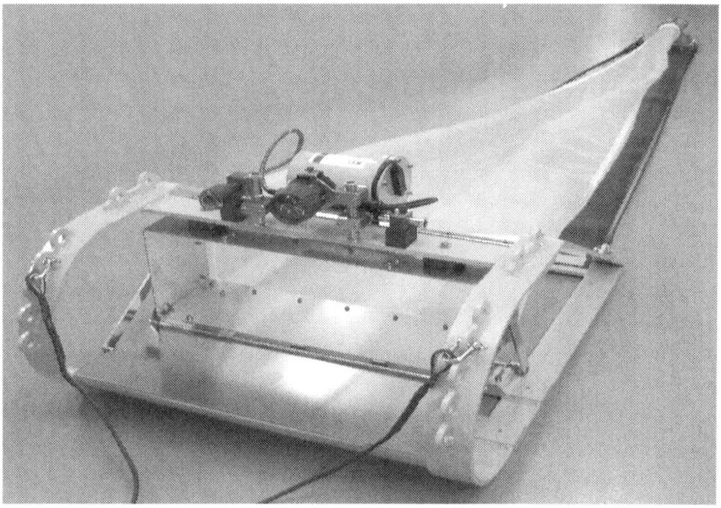

Courtesy of Sohier (2013)

The hyperbenthos net is used for the collection of solely hyperbenthos organism. The hyperbenthos sledge consist of the benthic carriage that drags a net over the sea floor. This net is divided into various compartments in a stacking order to allow the benthic communities to be collected throughout various level until up to the sea floor level.

TOOLS AND TECHNIQUES IN SAMPLING FOR BENTHIC ORGANISMS

The choice of tools selected for sampling seabed substrata or other benthic macro fauna can be determined by the hardness/ compactness of the substrata itself. Whilst a wide variety of sampling tools are available but only a small proportion of these tools have the ability to effectively collect samples from difficult areas. In these circumstances, researchers can use more than one tools to sample the full range of benthic organisms present in the area of sampling.

Hammon grab

Courtesy of Sohier (2013)

The Hamon grab is the recommended tool for sampling the benthic macro-fauna from coarse substrata. This tool consists of a rectangular frame that creates a stable support for a sampling bucket to attached on a pivoted arm. While reaching the seabed, the tension in the wire is released and activates the grab and moves the pivoted arm through a rotation of 90°. This cause the sample bucket through the sediment and seal itself by its rubber-covered steel plate in an inclined position. This will allow the sediment to roll easily towards the bottom of the sample bucket, reduce the risk of gravel becoming trapped between the leading edge of the bucket and retained the sample and thus preventing part of the sample being washed out. Some weights are attached to the grab to minimize the lateral movement of the supporting frame during sample collection. A drawback of the Hamon grab is that the sediment sample is mixed during the process of collection and retrieval.

Van Veen grab

Courtesy of Sohier (2013)

Researcher use Van Veen grab for benthic macro fauna studies. The Van Veen grab relies on the closure of two opposing jaws for the collection of a sediment sample. The Van Veen grab has long arms attached to each bucket, thus giving better leverage during closure. This mode of action is not ideally suited for the collection of coarse sediments as large particles of gravel tend to become caught between the jaws, resulting in loss of the sample upon retrieval of the grab.

Bottom Trawl

Researchers use bottom trawls for remotely sampling the epifauna. They are specifically designed to sample them just above the surface of the seabed. They are suitable for collecting larger, rarer or more motile species because bottom trawl relatively can cover large area in one deployment. The bottom trawl is relatively simple, it consists of a mechanism for keeping the mouth of the net open in horizontal and vertical dimensions and a net which guides fish inwards and a suitable mesh size, where the fish are collected. The size and design of net used is determined by the species being targeted, the engine power and design of the fishing vessel which through locally enforced regulations.

Beam Trawling

Courtesy of Sohier (2013)

Beam trawling is mainly used on smaller vessels, fishing for flatfish or prawns or those targeted sample that are relatively close inshore. The structure of beam trawling is that the mouth of the net is held open by a solid metal beam. It attached with two solid metal plate "shoes" which welded to the ends of the beam. It will slide over and disturb the seabed during sampling.

Otter trawls

Courtesy of Sohier (2013)

Otter trawls is the type of net that keeps the mouth of the net open. These trawls are commonly used to collect organisms from a sandy bottom.

When otter trawls pulled along the seabed, hydrodynamic pressure pushes them outwards, preventing the mouth of the net from closing. They also sometimes act like a plough and dig into the seabed. This creates a turbid cloud that scares the fish and allow them to move towards the opened trawl net. The net is held open vertically on an otter trawl by a float and/or by kites that attached to the "headline" and weighted "bobbins" by attached to the "foot rope". These bobbins vary in their design depending on the roughness of the sea bed, small rubber discs are used for very smooth, sandy ground and large metal balls are used for very rough ground. These bobbins can also be used for emergency as it was designed to lift the net off the seabed when they hit an obstacle.

Shrimp trawling

Courtesy of Sohier (2013)

Shrimp trawling is a special small-mesh bottom trawl that is used to catch northern shrimp. This follows the basic design of the otter trawl, but with modified shape and dimensions. The ground rope and sweep are configured to optimize the capture of shrimp.

Dredges

Dredges are used when there is a hard or uneven nature of the substrata that precludes the use of a trawl. In general, the use of towed dredges are for evaluation of epifaunal community structure.

Newhaven Scallop dredge

Courtesy of Sohier (2013)

The Newhaven Scallop dredge is a commercially used towed device that operated over very coarse terrain. The dredge itself consists of a triangular steel frame supporting on its underside, a spring-loaded plate to which a tooth bar, designed to dig into the sediment is bolted. When the dredge encounters rock or large stones, the springs allow the tooth-bar to swing back thus avoiding snagging and reducing the quantity of stones caught. A bag whose lower surface is made up of heavy-duty metal links with an upper surface of heavy gauge nylon mesh is attached to each frame. A number of these dredges are attached to a robust metal beam which is fitted with large rubber rollers at each end. The maximum diameter of particle likely to be retained within the dredge is approximately 20 mm.

The dredges are deployed over the stern or side of a vessel and towed for a pre-determined time. Care must be taken to ensure that the dredge is deployed the right way up. It will encounter damaged if towed over bedrock or through large boulders. The sampling efficiency of the dredge for each tow can be assessed on deck by quantifying the collected material. The use of this device is recommended for the collection of qualitative samples in areas of coarse and unconsolidated sediments which are too rough or uneven to permit the deployment of less robust gear. The Scallop dredge may be used to test the suitability of the ground prior to the usage of less robust gear such as beam trawl. This may be particularly useful if the ground is thought to be very coarse or uneven.

Rallier-du-bathy dredge

Courtesy of Sohier (2013)

The Raillier-du-Baty dredge is designed to operate in a range of substrata which is from sands to cobbles. It consists of a robust metal ring attached to a central towing arm. An open-ended bag of the desired mesh size is attached to the ring and the trailing end of the bag is tied to prevent loss of material during collection of the sample. The inner bag is protected by an outer, coarser bag which enclosed by a heavy-duty apron of fishing net, in order to reduce chafing. The warp is attached to a fixing point on the metal ring and a weak link is placed between this point and the central arm. This optimizes the digging capability of the edge of the ring and reduces the chances of the edge being lifted away from the seabed.

Corers

Corers work by boring a large tube into the benthos and then bringing up a column, or core of sediment intact within the tube. Caps can automatically seal off the ends of the core after it has pulled up a sample. This allow in protecting the sample and keeping it intact. Different sizes of cores and approaches work efficiently with different organisms and sediment types.

The gravity corer

Courtesy of U.S. Geological Survey

Gravity corers are widely used for the collection of the smallest marine metazoans from subtidal grounds. The gravity corer is basically a weighted tube mounted within a frame that descends by gravity from the research vessel up to the sea floor. It penetrates the sediment to a given depth, filling the tube with sediment in the process. The hydraulically-damped gravity corer has a slow rate of penetration that is controlled by a water-filled piston. Disturbance of the water-sediment interface can be either minimal or maximum. A core-catcher on the bottom of the tube moves into place when retrieval begins, trapping the sediment sample in the corner. A winch present onboard the ship will slowly bring the corer back to the surface.

Multiple corer

To test local faunal variations, it is necessary to recover several discrete samples from a single station. This can be done by multiple casts sampling with a single core tube, but this approach requires much shipping time. Therefore, with multiple corer it incorporates separate coring tubes into a single core body. It consists of a system with a series of tubes measuring about 4 cm in diameter. Above the system, a weight is mounted and this falls down onto the assembly system when the multicorer touches the sediment. The falling weight drives the tubes into the sea bed so that when they are raised again each of them contains a drilling core with sediment from the sea floor.

Courtesy of Sohier (2013)

Reineck boxcorer

Courtesy of Sohier (2013)

The boxcorer sampled relatively undisturbed samples. The equipment operates by a self-releasing trigger system that triggered by the frame when touching the sea bed. The square box is pushed into the bottom of the seabed by the gravity force of the weight that mounted on the top of the box retainer. A spade freed by the trigger-mechanism closes the sample box during the recovery of the unit. This is to prevent the sample from being washed-out.

Multi boxcorer

Courtesy of Sohier (2013)

A multi boxcorer incorporates the separate coring tubes of a boxcorer into a single core body for an efficient sampling.

TOOLS AND TECHNIQUES IN ANIMAL SAMPLING

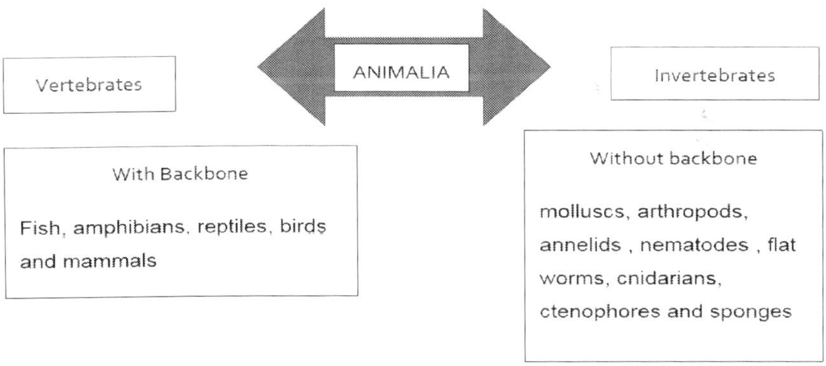

Again, the selection of tools always depends on the purposes of sampling and type of animals.

Always ask yourself these questions???? Why are you sampling? What are you sampling? How are you going to sample it out?

Are you going to bring the sample back to the laboratory or intend to carry out analysis in the field? For animal sampling, you also have to consider type of its movement (fast, moderate and slow) and the tools selection are depending on the area of specialization?

Pitfall Traps

Are often used to sample the small invertebrates living on the ground. This tool will help you to trap beetles and other insects, as well as spiders and slugs. The top should be level with the soil surface. Cover the trap with a stone or piece of wood to keep out the rain, to make it dark and to stop birds eating your catch. The traps must be checked often to avoid the animals escaping or being eaten before they are counted. As with most methods a large number of traps makes results more reliable and minimizes the effects of unusual results.

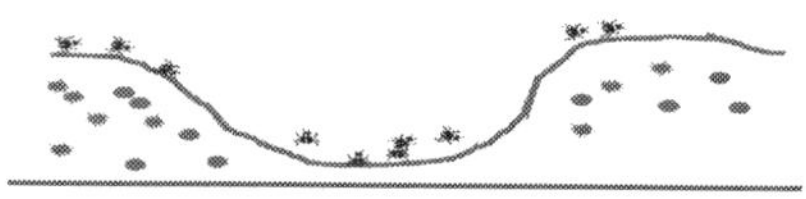

CHAPTER

3 TOOLS AND TECHNIQUES IN STORAGE BIOLOGICAL SAMPLE

INTRODUCTION

There are varieties of storage kit that one can use to stored collected sample safely for future uses. There is line of ground breaking products in all areas of storage from short term storage to long term storage. This storage will make anyone job in the laboratory easier and more efficient. Each consumable has their own advantages and function. Each of them offers useful tools for everyday laboratory work. If the specimen is small enough, include the entire plant with larger storage vial. Be sure to include samples of areas of healthy to diseased with different storage container. In this chapter, various storage tools that are mainly common in any type of laboratory was discussed thoroughly, how and when to use it as a storage tools.

PETRI DISHES

Petri dishes are shallow dishes with lids that are made from either plastic or glass. They are used to culture cells, preserve or grow samples or observe microorganisms. Petri dishes can be used to place the insect samples. Fill the petri dishes with water if the sample needs it for its survival. If the insect is moving too fast in the dish, place the insect in a refrigerator until it slows down. This generally takes from one to twelve hours.

JARS AND VIAL

Jars and vial are used to stored bigger biological samples such as leaves, roots and stem. Researcher that wanted to stored aquatic microorganisms, studying water chemistry or most likely those who collect water samples will need standard sized bottles or containers as below. It is all can be easily purchased with any scientific company that selling laboratory consumables.

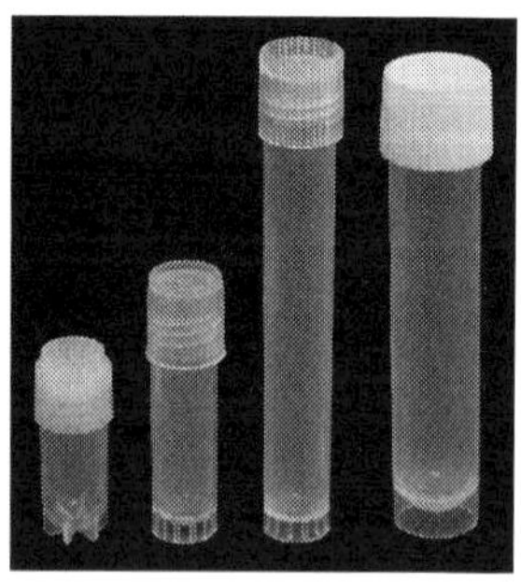

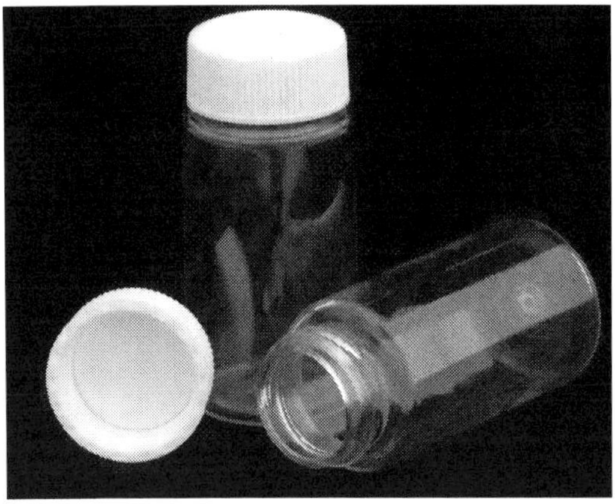

DESICCATOR

It is used to preserve the collected sample for subsequent analysis. It will also protect the material from air contamination or humidity.

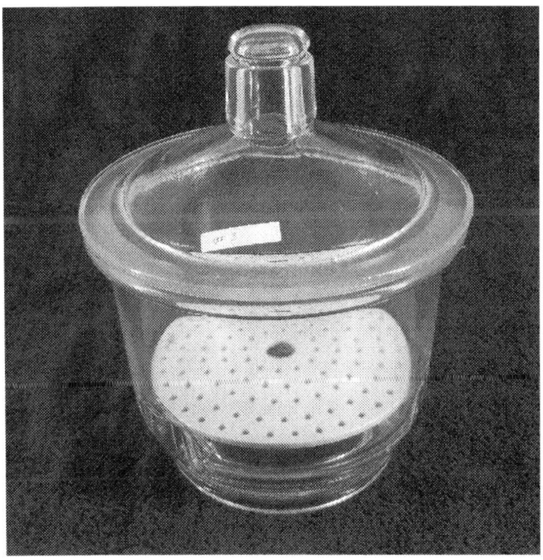

NEWSPAPER

To keep the sample dry until back to laboratory for example plant sample.

PLASTIC BAG

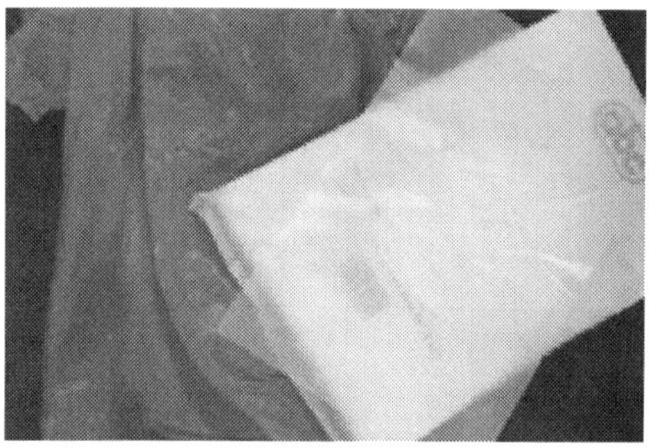

You may need heavy duty plastic back to carry certain sample or if you are transporting big volume of samples such as soil in the bag.

ALUMINUM FOIL

It depends what you are sampling. Just be sure to shake excess soil from the roots and wrap the plant in a manner that prevents soil from contaminating the foliage.

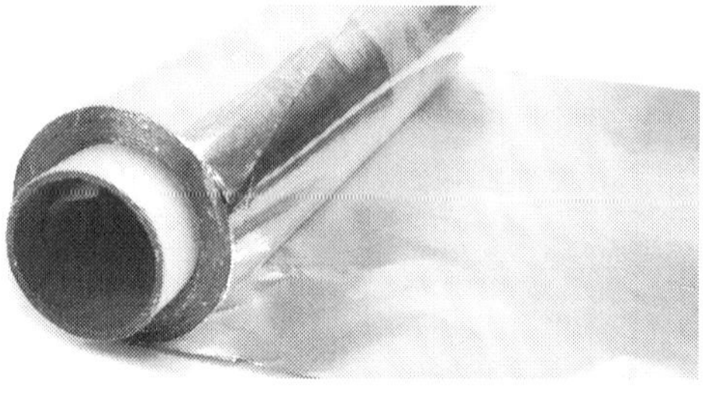

BOTTLE

Usually bottles are used for store reagents or samples.

INCUBATOR

Incubators are primarily used by microbiologists to control any necessary environmental factors such as temperature and humidity that may impact the samples that the microbiologists are growing and studying.

CRYOPRESERVATION

Cryopreservation is a process of cooling a sample to an extreme low temperatures at −80 °C using solid carbon dioxide or at −196°C using liquid nitrogen. At low temperatures, enzymatic or chemical activity which have potential to damage the biological material can be stopped. Once the preserved material has become frozen, it is relatively safe from further damage. In addition, cryopreservation methods will avoid damage caused by the formation of ice during normal freezing. Traditional cryopreservation has relied on coating the material with cryo protectant prior to freezing. However, to date usage of cryoprotectant are not a preferable choice. A maximum storage period of 1000 years were estimated based on the accumulation of radiation-induced DNA damage during cryonic storage.

Process of cryopreservation:

- As ice crystals grow in freezing water, solutes are excluded, causing them to become concentrated in the remaining liquid water. High concentrations of some solutes can be very damaging.

- Extracellular ice formation occurred when tissues are cooled slowly, water migrates out of cells and ice forms in the extracellular space. Too much extracellular ice can cause mechanical damage to the cell membrane due to crushing. Intracellular ice formation is also almost fatal to cells.
- Migration of water can also cause cellular dehydration. This stresses can damage on the cell directly.

A. Types of Freezing: Slow programmable freezing

Controlled-rate and slow freezing, also known as *slow programmable freezing (SPF)* enabled freezing the biological samples by using programmable programs, or by controlled rates. Such machines are used for freezing oocytes, skin, blood products, embryo, sperm, stem cells and general tissue preservation in hospitals, veterinary practices and research laboratories around the world.

Lethal intracellular freezing can be avoided if cooling is slow enough to permit sufficient water to leave the cell during progressive freezing of the extracellular fluid. To minimize the growth of extracellular ice crystal and recrystallization, biomaterials such as alginates, poly vinyl alcohol or chitosan can be used to impede the ice crystal growth along with traditional small molecule cryoprotectants. That rate differs between cells sizes and water permeability. For example a typical cooling rate of about 1 °C/minute is appropriate for many mammalian cells after treatment with cryoprotectants such as glycerol or dimethyl sulphoxide. The 1 °C / minute rate can be achieved by using devices such as a rate-controlled freezer or a benchtop portable freezing container.

Several independent studies have provided evidence that frozen embryos stored using slow-freezing techniques may in some ways be 'better' than fresh in IVF. The studies indicate that using frozen embryos and eggs rather than fresh embryos and eggs reduced the risk of stillbirth and premature delivery

B. Vitrification

Vitrification provides the benefits of cryopreservation without damage cause by ice crystal formation. The situation became more complex with the development of tissue engineering as both cells and biomaterials need to remain ice-free to preserve high cell viability and functions, integrity of constructs and structure of biomaterials. Vitrification of water is promoted by rapid cooling and can also be achieved without cryoprotectants but only with an extremely rapid decrease of temperature. For clinical cryopreservation, vitrification usually requires the addition of cryoprotectants prior to cooling. The cryoprotectants act like antifreeze: they decrease the freezing temperature and increase the viscosity. Instead of crystallizing, the syrupy solution becomes an amorphous ice and vitrifies. However certain sugars do not readily permeate through the membrane. Those solutes that do, such as dimethyl sulfoxide, a common cryoprotectant, are often toxic in intense concentration. Rather than a phase change from liquid to solid by crystallization, the amorphous state is like a "solid liquid"and the transformation is over a small temperature range described as the "glass transition" temperature. One of the difficult compromises of vitrifying cryopreservation concerns limiting the damage produced by the cryoprotectant itself due to cryoprotectant toxicity.

Freeze-able Tissues

Freezing tissues are much easier than organ due to its thin layers and small clumps of individual cells. This type of sample can be cooled more quickly and require lesser doses of toxic cryoprotectants. Suitable combinations of cryoprotectants and regimes of cooling and rinsing during warming often allow the successful cryopreservation of biological materials.

Embryos

- Cryopreservation for embryos is used for embryo storage, for example in vitro fertilization (IVF).

Ovarian tissue

- Cryopreservation of ovarian tissue is of interest to women who want to preserve their reproductive function beyond the natural limit, or whose reproductive potential is threatened by cancer therapy. The procedure is to take a part of the ovary tissue and perform slow freezing before storing it in liquid nitrogen whilst therapy is undertaken. Tissue can then be thawed and implanted near the fallopian, either orthotopic (on the natural location) or heterotopic (on the abdominal wall), where it starts to produce new eggs, allowing normal conception to occur.

Oocytes

- Human *Oocyte cryopreservation* is a new technology in which a woman's eggs (oocytes) are extracted, frozen and stored. Later, when she is ready to become pregnant, the eggs can be thawed, fertilized and transferred to the uterus as embryos.

Semen

- Semen can be used successfully almost indefinitely after cryopreservation. It can be used for sperm donation where the recipient wants the treatment in a different time or place, or as a means of preserving fertility for men undergoing vasectomy or treatments that may compromise their fertility, such as chemotherapy, radiation therapy or surgery.

Moss

- Cryopreservation of whole moss plants are also achievable. Preservation of microbiology cultures for short term can be done through refrigerator. However, in this condition, cell division and metabolism is not completely stopped. For long term storage to preserve cultures genetically or phenotypically, as cell divisions can lead to mutations or sub-culturing which can cause phenotypic changes. A preferred option is cryopreservation.

Fungi

- Fungi, notably zygomycetes, ascomycetes and higher basidiomycetes, regardless of sporulation, are able to be stored in liquid nitrogen or deep-frozen. Cryopreservation is a hallmark method for fungi that do not sporulate, sporulate with delicate spores, pathogenic and to be used for genetic stocks. As with many other organisms, cryoprotectants like DMSO or glycerol are used. Differences between choosing cryoprotectants are based on species dependent. For penetrating fungi, cryoprotectants like DMSO, glycerol or polyethylene glycol are most effective than non-penetrating ones include sugars, mannitol, sorbitol, dextran. Freeze-thaw repetition is not recommended as it can decrease viability.

Multiple protocols for freezing are summarized below (each uses screw-cap polypropylene cryotubes)

- For non-sporulating fungi or embedded mycelia, use 10% glycerol. It is added into agar plugs of fresh culture and immediately frozen into liquid-nitrogen vapor (−170 °C).
- For spores or mycelia from agar plate, use 10% glycerol or 5% DMSO spore for mycelia suspension.
- For liquid mycelia, mycelia are macerated and mixed to make a final concentration of 10% glycerol or 5% DSMO.

Bacteria

Many common culturable laboratory strains are deep-frozen to preserve genetically and phenotypically stable for long-term stocks. Sub-culturing and prolonged refrigerated samples may lead to loss of plasmid(s) or mutations. Common final glycerol percentages are 15, 20 and 25. From a fresh culture plate, one single colony of interest is chosen and multiplied in liquid culture. From the liquid culture, the medium is directly mixed with equal amount of glycerol. Mixing of the solution need to be done gently by inversion and cooling can be done by placing the cryotube directly at −50 to −95 °C. Shock-freezing in liquid nitrogen or gradually cooling and then storing at −80 °C or cooler (liquid nitrogen or liquid nitrogen vapor). Recovery of bacteria can also vary, namely if beads are stored within the tube then the few beads can be used to plate or the frozen stock can be scraped with a loop and then plated, however, since only little stock is needed the entire tube should never be completely thawed and repeated freeze-thaw should be avoided. 100% recovery is not feasible.

VIABILITY ASSAY

A viability assay is an assay to determine the ability of organs, cells or tissues to maintain or recover viability. Viability can be distinguished from the all-or-nothing states of life and death by use of a quantifiable index between 0 and 1 (or 0% and 100%). Viability can assay mechanical activity, motility (spermatozoa or granulocytes), contraction (muscle tissue or cells), mitotic activity, etc.

For example, examining the ratio of potassium to sodium in cells can serve as an index of viability. If the cells do not have high intracellular potassium and low intracellular sodium, then (1) the cell membrane may not be intact and/or (2) the sodium-potassium pump may not be operating well. As with many kinds of viability assays, quantitative measures of physiological function do not indicate whether damage repair and recovery is possible. An assay of the ability of a cell line to adhere and divide may be more indicative of incipient damage than membrane integrity. Fluorescent-based assays do not require large sample sizes. Viability assays are used to assess the success of cryopreservation techniques, the toxicity of substances, or the effectiveness of substances in mitigating effects of toxic substances.

1. **Cytolysis or membrane leakage assays**: This category includes the lactate dehydrogenase assay, a stable enzyme common in all cells which can be readily detected when cell membranes are no longer intact. Examples:

   Propidium iodide

   Trypan blue

   7-Aminoactinomycin D

2. **Mitochondrial activity or caspase assays**: Resazurin and Formazan (MTT/XTT) can assay for various stages in the apoptosis process that foreshadow cell death.

3. **Functional assays**: Assays of cell function will be highly specific to the types of cells being assayed. For example, motility is a widely used assay of sperm cell function. Gamete survival can be used to assay fertility, in general. Red blood cells have been assayed in terms of deformability, osmotic fragility, hemolysis, ATP level and hemoglobin content. For transplantable whole organs the ultimate assay is the ability to sustain life after transplantation, an assay which is not helpful in preventing transplantation of non-functional organs.

4. **Genomic and proteomic assays**: Cells can be assayed for activation of stress pathways using DNA microarrays and protein chips.

CHAPTER

4 TOOLS AND TECHNIQUES IN BIOLOGICAL SAMPLE PROCESSING

INTRODUCTION

In order to process collected sample various buffers and solution are need to be prepared. Below are some comman buffers and solution that common in sample processing.

COMMON SOLUTION RECIPES

Ampicilin (50 mg/mL)

Dissolve 0.5 g of ampicilin in 10 mL of distilled water. Filter sterilised the content prior storing at -20 °C.

Spectinomycin (50 mg/mL)

Dissolve 0.5 g of spectinomycin in 10 mL of distilled water. Filter sterilised the content prior storing at -20 °C.

Rifampicin (25 mg/mL)

Dissolve 0.25 g of spectinomycin in 10 mL of distilled water. Filter sterilised the content prior storing at -20 °C.

IPTG (0.1 mM)

Dissolve 0.2 mg of IPTG (MW 238.3) in 1 mL distilled water. Filter sterilised the solution prior storing at -20°C.

X-gal (60 µg/mL)

Dissolve 0.6 g of X-gal in 10 mL of DMSO (dimethyl sulfoxide). Filter sterilised the solution. Wrapped the vial with aluminium foil prior storing it at -20°C.

Acetosyringone (100 mM)

Dissolve 0.98 g of acetosyringone (MW 196.20) in 50 mL methanol.

5-bromo-4-chloro-3-indoyl-β-D-glucuronide (X-Gluc) (200 mM)

Dissolve 2 mg of X-Gluc (MW 521.79) in DMSO prior storing it at -20 °C.

COMMON MEDIA PREPARATION

LB agar

Dissolve 37 g of the medium thoroughly in 1 L of distilled water. Autoclave the medium at 121°C for 15 minutes before pouring it in a petri dishes.

LB Broth

Dissolve 25 g of the medium in 1 L of distilled water. Autoclave the medium at 121°C for 15 minutes before substantial usage.

LB agar 50 µg/L ampicilin

LB agar solution	1 L	
50 µg/L ampicilin	50mg/mL	1µL

LB broth 50 µg/L ampicilin

LB broth solution		1 L
50 µg/L ampicilin	50mg/mL	1µL

LB agar 50 mg/L spectinomycin

LB agar solution		1 L
50 mg/L spectinomycin	50mg/mL	1mL

LB broth 50 mg/L spectinomycin

LB broth solution		1 L
50 mg/L spectinomycin	50mg/mL	1mL

LB agar 50 mg/L spectinomycin and 25 mg/L rifampicin

LB agar solution		1 L
50 mg/L spectinomycin	50mg/mL	1mL
25 mg/L rifampicin	25mg/mL	2mL

LB broth 50 mg/L spectinomycin and 25 mgLl rifampicin

LB broth solution		1 L
50 mg/L spectinomycin	50mg/mL	1mL
25 mg/L rifampicin	25mg/mL	2mL

LB agar 25 mg/L rifampicin

LB agar solution		1 L
25 mg/L rifampicin	25mg/mL	2mL

LB broth 25 mg/L rifampicin

LB agar solution		1 L
25 mg/L rifampicin	25mg/mL	2mL

0.8% and 1% agarose gel

For 0.8% gel, dissolve 0.4 g of agarose gel in 50 mL of distilled water. For 1 % gel, dissolve 0.5 g of agarose gel in 50 mL of distilled water. Use microwave to dissolved the agar.

COMMON BUFFERS PREPARATION

Tris-HCL (1 M pH 8.0 50 mL)

Dissolve 6.1 g of Tris (MW 121.1) in distilled water of ½ of the desired volume. Add 1M HCl until the pH meter indicated 8.0. Add distilled water to top up to the remaining volume.

EDTA (1 M pH 8.0)

Dissolve 9.3 g of EDTA (MW 372.24) in ½ of the desired volume. Add NaOH pellet until the pH meter indicated 8.0. Add distilled water to top up to the remaining volume.

NaCl (5 M)

Dissolve 14.61 g of NaCl (MW 58.44) in 50 mL of distilled water.

RNA extraction buffer (50 mL)

0.25 M Tris-HCL pH 8.0	1 M	12.5 mL
0.05 M EDTA	1 M	2.50 mL
2.50 M NaCl	5.0 M	25.0 mL

2% CTAB	1.00 g
3% PVP-40	1.50 g
5% mercaptoethanol	2.50 mL
Deionised water	to 50 mL

Choloroform:Isoamylalcohol (24:1)

chloroform	24 mL
isoamylalcohol	1 mL

Sodium acetate (3 M pH 5.2)

Dissolve 40.8 g of sodium acetate (MW 82.03) in 80 mL of distilled water, add glacial acetic acid until the pH solution reached 5.2. The remaining solution was top up to 100 mL.

Pyrogallol (1%)

Dissolve 1 g of pyrogallol in 100 mL of ethanol. Pyrogallol solution are prepared freshly every application as it is easily oxidized if stored for long. Prepared pyrogallol are kept in dark bottle throughout the usage.

Soil composition

1:2:1 consist of soil: peatgro: sand

Phenol:Chloroform 1:1

Phenol	25 mL
Chloroform	25 mL
Volume	up to 50 mL

TAE Buffers (1 L of 1X)

1X TAE buffers	10x	100 mL
Distilled water	up to 1L	

$MgSO_4$ (1 M)

Dissolve 12 g of $MgSO_4$ (MW 120.38) in 100 mL of distilled water.

MES (0.5 M)

Dissolve 9.76 g of MES in 100 mL of distilled water.

Induction medium (1 L of pH 5.6)

K_2HPO_4		10.5 g/L
KH_2PO_4		4.5 g/L
$(NH_4)_2SO_4$		1 g/L
NaCitrate		0.5 g/L
glucose		1 g/L
fructose		1 g/L
glycerol		4 g/L
1 mM $MgSO_4$	1.0 M	1 µL
10 mM MES	0.5 M	20 µL
Distilled water		up to 1 L

*pH adjustment are carried out in ½ of the desired volume by using NaOH solution.

Infiltration medium (1 L of pH 5.6)

10 mM $MgSO_4$	1.0 M	10 µL
10 mM MES	5 M	20 µL
Distilled water		up to 1L

*pH adjustment are carried out in ½ of the desired volume by using NaOH solution.

Acetosyringone (100 µM in 5 mL induction medium)

100 µM acetosyringone	100 mM	5 µL

Acetosyringone (200 µM in 5 mL infiltration medium)

200 µM acetosyringone	100 mM	10 µL

Sodium phosphate (0.2 M pH7.0)

Consists of NaH_2PO_4 and Na_2HPO_4. $NaH_2PO_4.H_2O$ (MW 138.01); 0.2 M solutions contains 27.6 g/L. $Na_2HPO_4.12H_2O$ (MW 358.22); 0.2 M solution contains 71.64 g/L

pH	x mL 0.2M- NaH_2PO_4	y mL 0.2M- Na_2HPO_4
7.0	30.5	19.5

GUS staining buffers (100 mL)

50 mM sodium phosphate pH7.0	0.2 M	25 mL
1 mM EDTA	1 M	0.1 mL
2 mM X-Glu	200 mM	1 mL
0.05% SDS		0.05 g
0.1% sodium N-lauryl-sarcosine		0.1 g
0.1% Triton X-100		0.1 mL
Distilled water		up to 100 mL

hygromycin in 0.05% tween 20 (150 mg/L)

Hygromycin	0.15 g
Tween 20	0.5 mL
Distilled water	up to 1L

NaOH (0.5 N of 50 mL)

Dissolve 1 g of NaOH (MW 40) in 50 mL of distilled water.

BASIC TOOLS IN BIOLOGICAL SAMPLE PROCESSING

Graduated cylinder

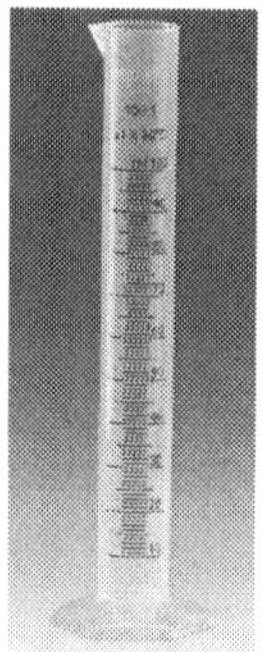

Used for measuring liquid with better accuracy than beaker

Beaker

Used for measuring liquid roughly volume with low accuracy. It is used to prepare buffers and solution. Though usually used by chemists, biologists also use beakers--a plastic or glass container to store solutions used in biological experiments. Beakers are also ideal to mix solutions.

Erlenmeyer flask (conical flask)

Used for titration or filtration of liquids and to prevent air contamination to the sample during work.

Test tube holder

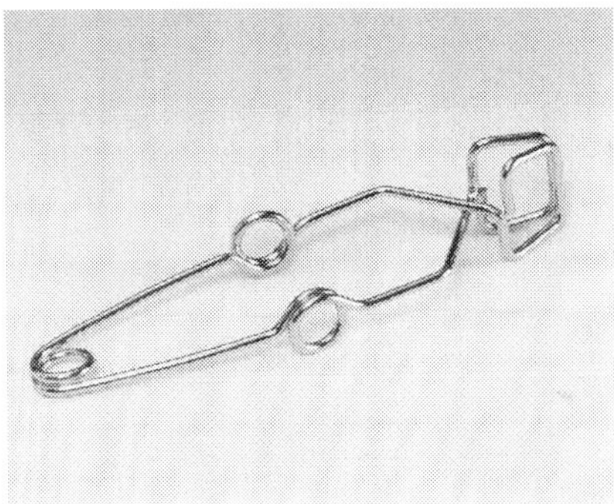

Used for holding hot test tubes.

Crucible

Used for burning sample at high temperature.

Mortar and Pestle

Used for grinding materials which have large particle to small size.

Round bottom flask

Used for distillation or heating of liquid, allows uniform heating.

Volumetric flask

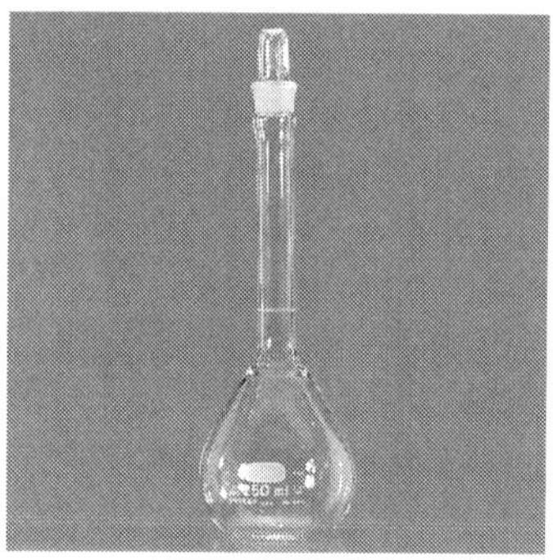

Used for measuring liquid with high accuracy. It is also used while preparing buffer and reagents solutions.

Separator funnel

Used for Liquid-Liquid extracts, designed for increase separation efficiency.

Filter paper

Used for filtration, available with different pore size.

Buchner funnel

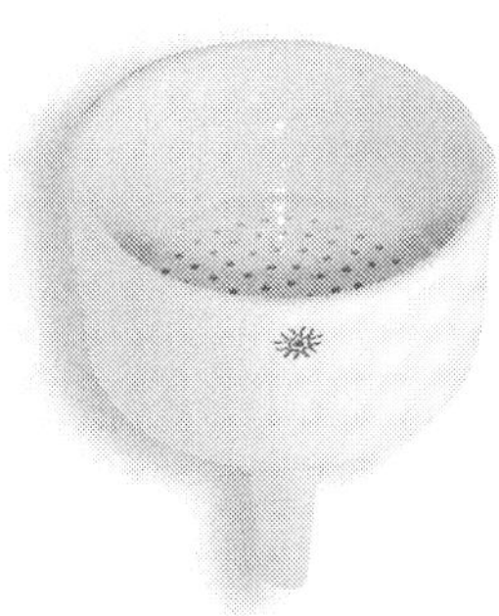

Used for vacuum filtration using filter paper.

Watch Glass

Used for air drying or oven drying of liquid.

Funnel

Used for liquid transfer. Also certain times can be use for sample filtration.

Wash bottle

Used for dispensing small amount of liquid like solvent of distilled water.

Bottle dropper

Used mainly for indicators or reagents.

Distillation head

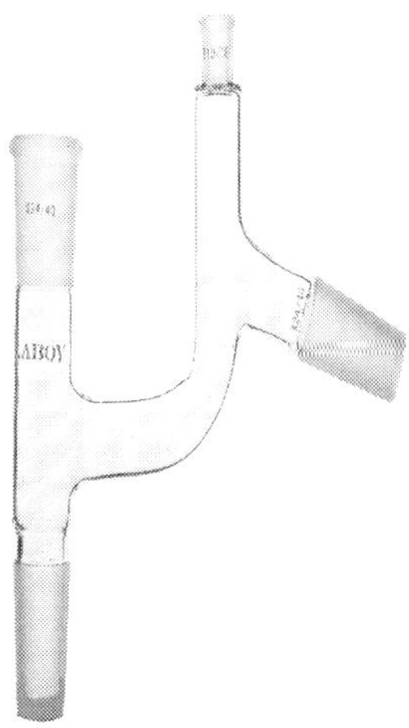

Used for connecting distillation flask with condenser and thermometer in the distillation system.

Receiving adapter

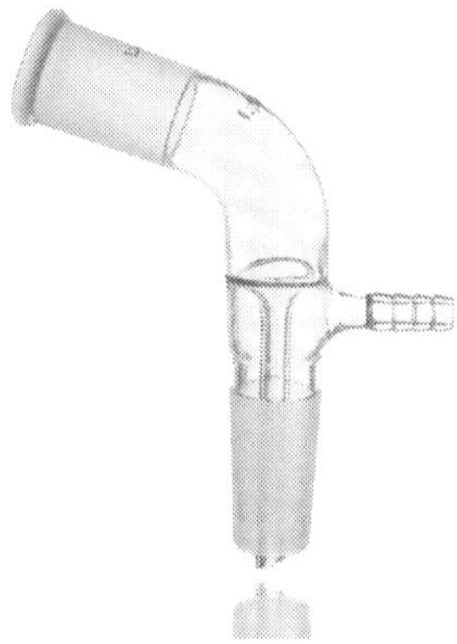

Used to connect condenser with receiving flask in the distillation system.

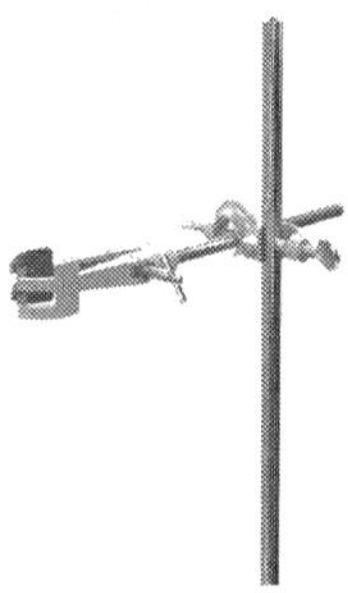

Used commonly as base for holding distillation system and burette along with clamp and boss head.

Ring with boss head

Used for holding separator funnel or funnel along with stand.

Bunsen burner

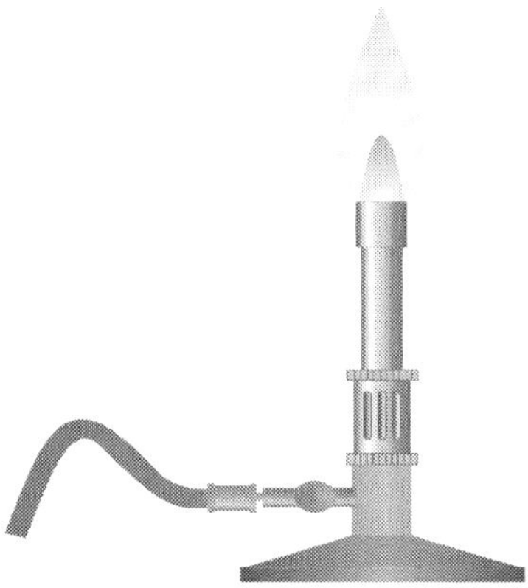

A Bunsen burner is essentially a tube connected to a steady supply of flammable gas that is used in the laboratory to provide an adjustable open flame. The flame is used to sterilize inoculating loops and needles. Bent glass rods used for spreading liquid on agar plates and forceps are also sterilized with the flame. However, the glass rods and forceps are not heated to red hot, but dipped in alcohol and then the alcohol is burned off after brief contact with the flame.

Open flame is a safety hazard and must not be left unattended. Make sure your working area is clear, organized and free from flammable materials. Be sure to remove any dangling or loose clothing and tie back long hair before using the Bunsen burner.

Following are the five steps for correctly using a Bunsen burner:

1. Obtain a striker and check to be sure that flint remains (black portion in video). To replace a worn flint, simply unscrew the old one and screw on a new one.
2. Listen carefully while you turn on the gas until you can just barely hear it flow. (If the gas is turned up too high, the flame will blow itself out.)
3. Immediately strike a spark over the Bunsen burner. The striker is designed for left-handed use. If you use your right hand, you may need to pull up with your thumb and push down with your fingers to obtain enough friction to cause a spark.
4. Adjust the air vents at the bottom of the Bunsen burner

to obtain a bright blue cone of flame inside the larger, darker flame. The hottest part of the flame is at the top of the bright blue cone.

5. Turn off the burner when it is not in use.

Dropper

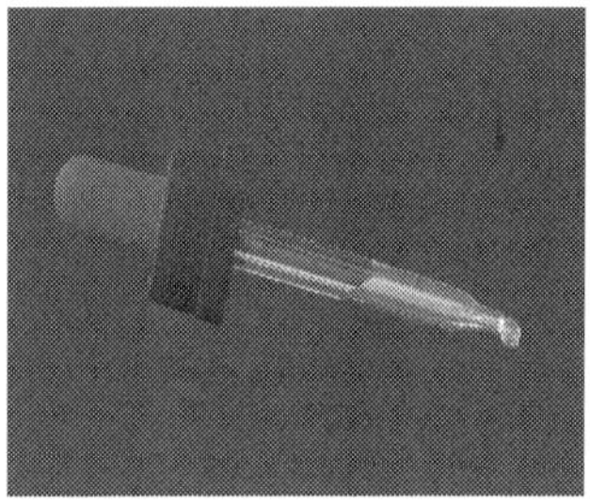

Used for transfer liquid drop by drop.

Burette

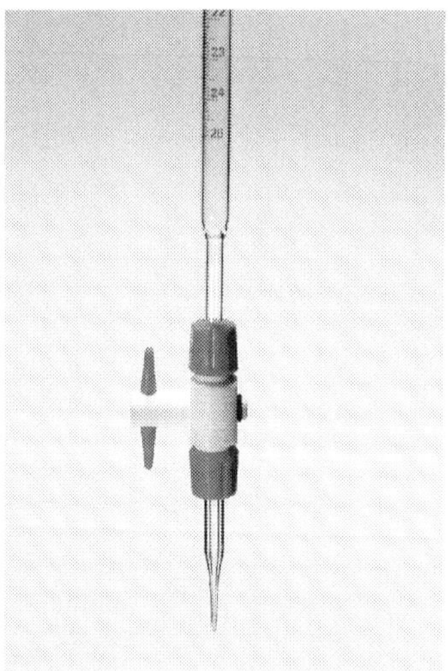

A burette is a long and thin cylindrical object with an open top and a stopcock at the bottom. Used for titration where titrant pored inside. Measurement labels run along the length of burets. Liquids are poured into the open top of burets and can be released through the stopcock at the bottom. Burets are used when a precise amount of liquid needs to be used.

Burette Clamp

Used along with stand to hold burette.

Bottle top dispenser

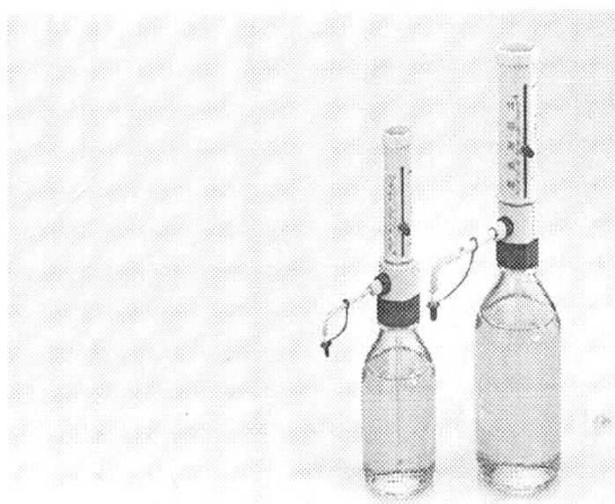

Used for transfer accurate amount from the bottle mostly used for acids and organic solvent.

Glass rode

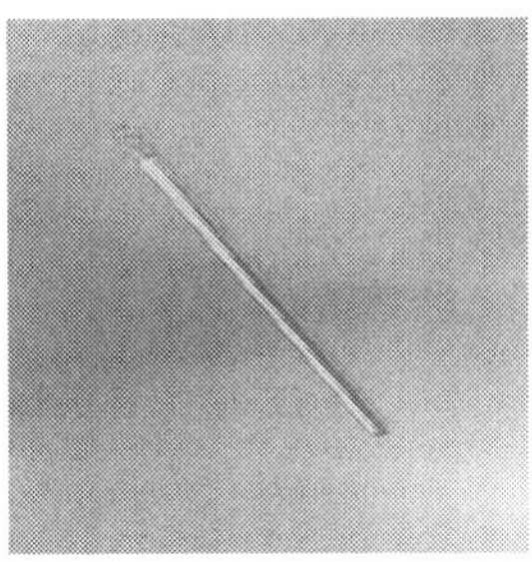

Used for stirring of liquid for several purposes.

Bruch

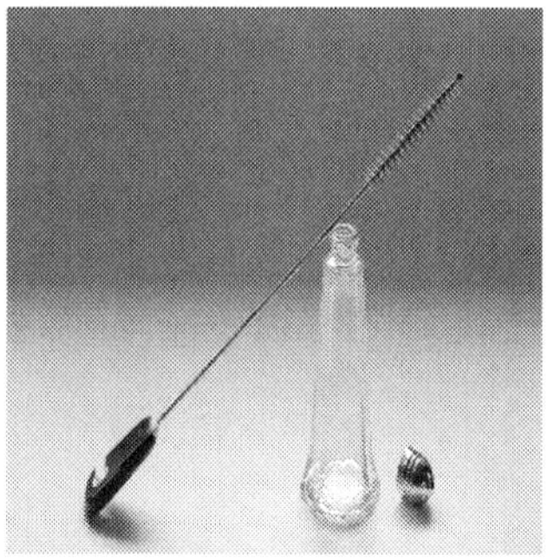

Used for cleaning of all types of glassware.

Boss head

Used for holding clamp with stand rode in the distillation system.

Thermometer

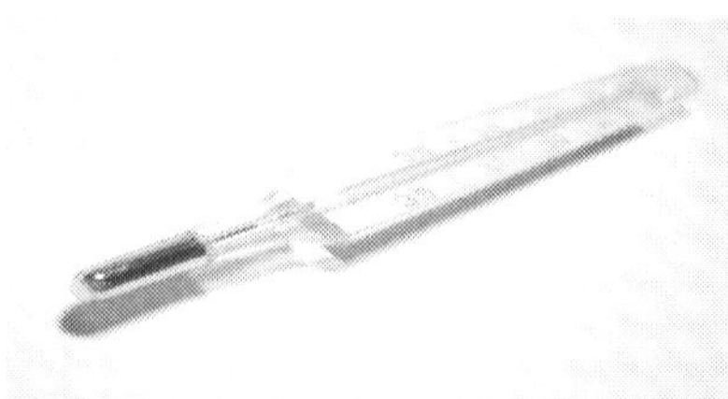

Used for measuring temperature.

Condenser

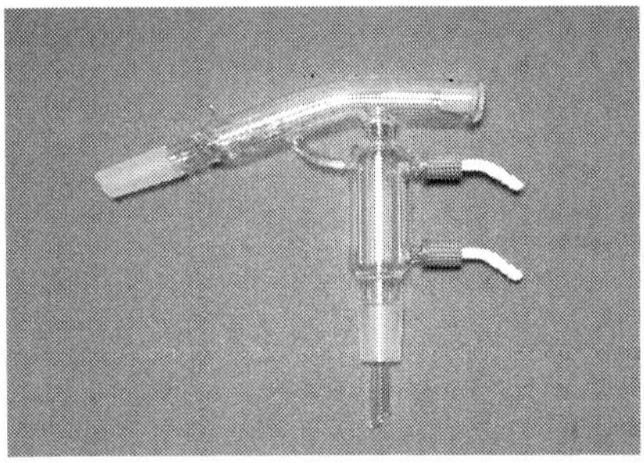

Used for condensing volatile liquid during distillation.

Clamp

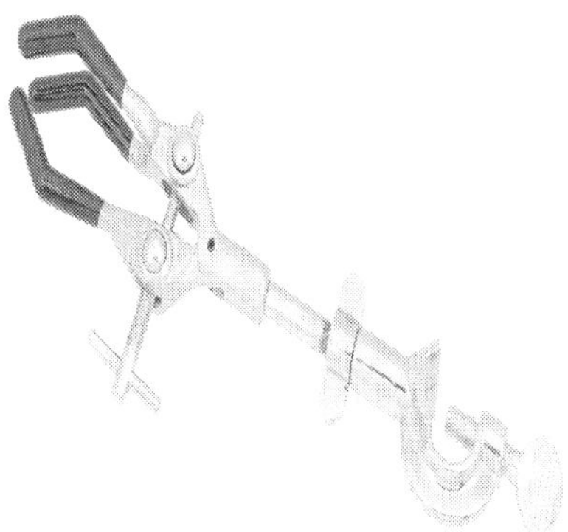

Used for holding glassware along with stand and boss head.

Wire gauze

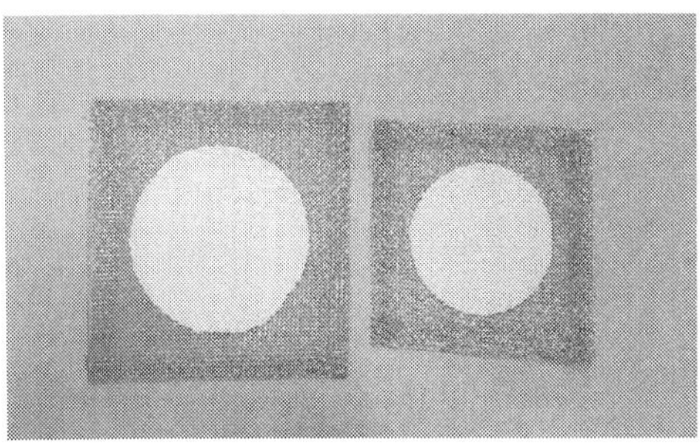

Used for spreading the head of burner homogeneously.

Forceps

Forceps look similar to tweezers. Forceps are primarily used to retrieve samples from organs or tissue, or to place samples on microscope slides.

Scalpel

Scalpels are small instruments that are often four to five inches in length and they have a sharp blade at one end. Scalpels are used during dissecting tissue or organs. Also can be used for holding small objects like filter paper or sample tissues.

Water Bath

Water bath is a laboratory equipment made from a container filled with heated water. It is used to incubate samples in water at a constant temperature over a long period of time. All water baths have a digital or an analogue interface to allow users to set a desired temperature. Water bath can be used for warming of reagents, melting of substrates or incubation of cell cultures. It is also used to enable certain chemical reactions to occur at high temperature. Water bath is a preferred heat source for heating flammable chemicals instead of an open flame to prevent ignition. There are different types of water baths which are designed based on application. For all water baths, it can be used up to 99.9 °C. When temperature is above 100 °C, alternative methods such as oil bath, silicone bath or sand bath may be used.

Circulating Water Bath

Circulating water bath (also called stirrers) are ideal for applications when temperature uniformity and consistency are critical, such as enzymatic and serologic experiments. Water is thoroughly circulated throughout the bath resulting in a more uniform temperature.

Non-Circulating Water Bath

This type of water bath relies primarily on convection instead of water being uniformly heated. Therefore, it is less accurate in terms of temperature control. In addition, there are add-ons that provide stirring to non-circulating water bath to create more uniform heat transfer.

Shaking Water Bath

This type of water bath has extra control for shaking, which moves liquids around. This shaking feature can be turned on or off. In microbiological practices, constant shaking allows liquid-grown cell cultures grown to constantly mix with the air.

Water Bath Controls

Temperature Control: All water bath have a control to set temperature. This control can be digital or manual. Often there is an indicator light associated with this control. When the light is on, the water bath is heating. When the water bath reaches the set temperature, it will cycle on and off to maintain constant temperature.

Safety Control: All water bath has a second control called the safety. This control is set at the maximum temperature the water bath should attain. It is usually set just above the temperature control. Often an indicator light is associated with the safety control. If the water bath reaches the temperature that the safety control is set at, the light will go on. It will be impossible for the water bath to heat higher than the safety setting even when the temperature setting is higher. If your water bath stays a temperature lower than the temperature control setting, try increasing the safety control setting.

Shaking Control: All water bath has additional controls for shaking. The shaking mechanism can be turned on or off. The speed of shaking can also be set. Cultures grown in liquid media are often shaken to allow constant mixing with air and oxygen with the culture.

Precautions

It is not recommended to use water bath with moisture sensitive or pyrophoric reactions. Do not heat a bath fluid above its flash point. Water level should be regularly monitored and filled with distilled water only to ensure cleanliness and sterility. This is required to prevent salts from depositing on the heater. Disinfectants can be added to prevent growth of organisms. Raise the temperature to 90 °C or higher once a week for half an hour for the purpose of decontamination. Use water resistant marker as markers can come out easily in water bath. If application involves liquids that give off fumes, it is recommended to operate water bath in fume hood or in a well-ventilated area. The cover is closed to prevent evaporation and to help reaching high temperatures. Set up on a steady surface away from flammable materials.

Heat Mantle

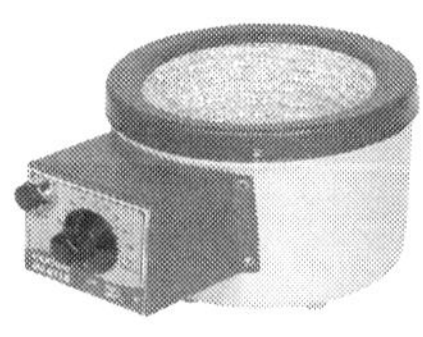

Used as a source of heat, mainly for distillation system.

Balance

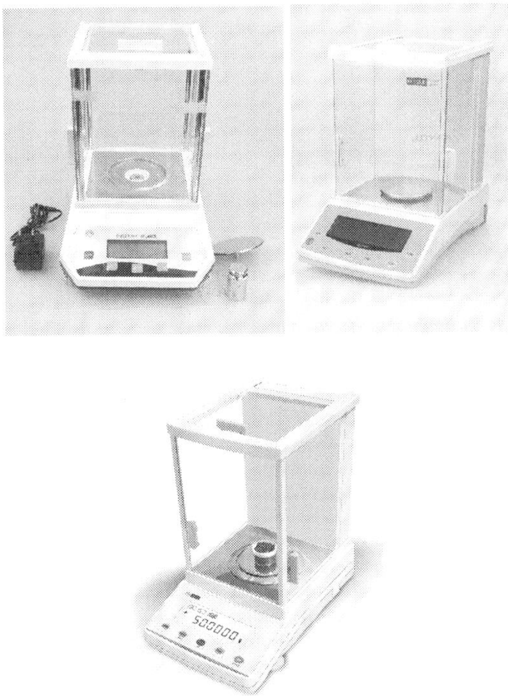

Figure: It above shows an electronic analytical balancer.

This balancer is used to weigh research materials. In order to use this balancer, the user need to tare this equipment to 0.0000g if it indicates something other than 0.0000g. Taring can be done by simply pressing the tare button. Due to its high sensitivity, an analytical can detect the mass of a single grain of a chemical substance. Thus, if a method of direct weighing is used, the substance that to be added is placed onto the tared container. The container used should be completely dry inside and out at room temperature. Even a slight temperature differences can produce significant changes in mass of the container.

For sample that is in a weighing bottle, the weighing bottle is tipped above the container to receive the sample and a small amount is allowed to fall out of the weighing bottle. Prior closing the weighing bottle, the weighing bottle is tipped back up and tapped gently to make sure all of the substance falls back in the bottle and doesn't remain on the bottle rim. The cap is replaced and the bottle weighed once again to ensure the current weight. The difference between the first and second weighing represents the amount transferred. If the sample has a tendency to absorb water and thus to gain weight when exposed to the moisture of the air, then the user need to minimize exposure to the atmosphere.

If a beaker were to be used the beaker must be clean without any fingerprints and particles sticking to the bottom of the beaker, dry and at the same temperature as the room and the balance. Even though, the mouth of the beaker is rather high, the user only needs a simple coordination to position the spatula over the mouth without knocking it against the beaker or the windows of the balance, as this will lead to losing the sample. However, beaker provides a bigger space for sample addition and the placement of your hands, arms and the transfer spatula is much easier. A spatula/fine spatula can be used to remove a sample from the weighing bottle and the sample placed directly into the weighing container. The disadvantage is that the user must check the weight of the beaker and the sample after each addition and it is not good practice to remove excess sample from the beaker back to the sample bottle. For big and dried material, user can still weigh the sample directly but she/he need to do in a good manner which doesn't spoiled the balancer and experiment materials.

If the user is using Erlenmeyer flask he/she might risk losing the sample outside the flask due the small diameter of the flask neck, the entry point is rather high. In order to use Erlenmeyer flask effectively the user need to have good position in orienting the spatula above the mouth of the flask. The advantage of using Erlenmeyer flask is that multiple transfer from weighing container which lead to loss of material can be minimize.

Using the Top Loading Balance

Several top loading balances are precise up to ±0.001 g and are available for use in the balance room. The top loading balance is the instrument of choice where precision is not so important. Similar to other balancer, top loading balance can also be tared by pressing the tare button. It is good to place a weighing paper or an aluminum foil on the pan prior to removing the sample from its container. However, this paper is not encouraged to be used in the analytical balances, because there is always the chance that some of the substance being weighed will stick to the weighing paper after the weight has been recorded thus producing an error on the low side.

Dissecting tools

To perform dissections, you need tools such as scalpels and tweezers.

Homogenizer

Homogenizer is used to homogenized the cell walls and cell membranes for getting the suspension of the cell component out from the cell. Usually the plant and animal cell is placed in hypotonic solution prior to homogenization.

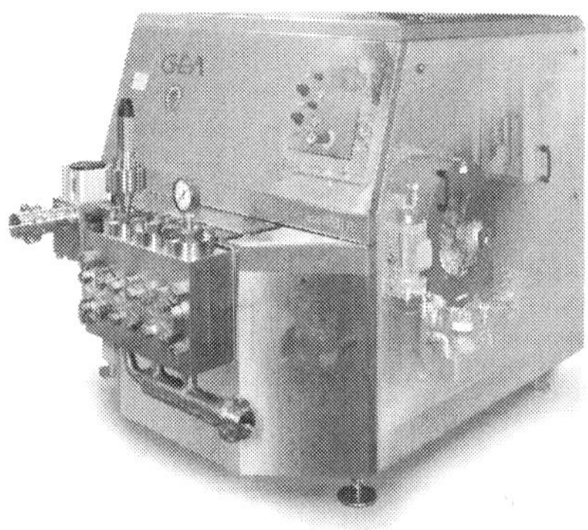

Centrifuge and Microcentrifuge Tube

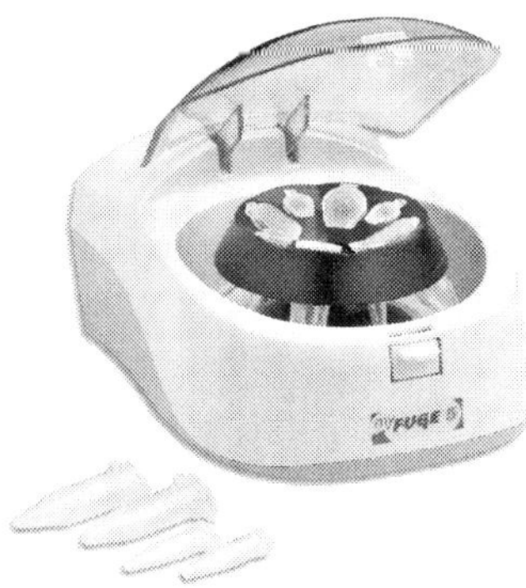

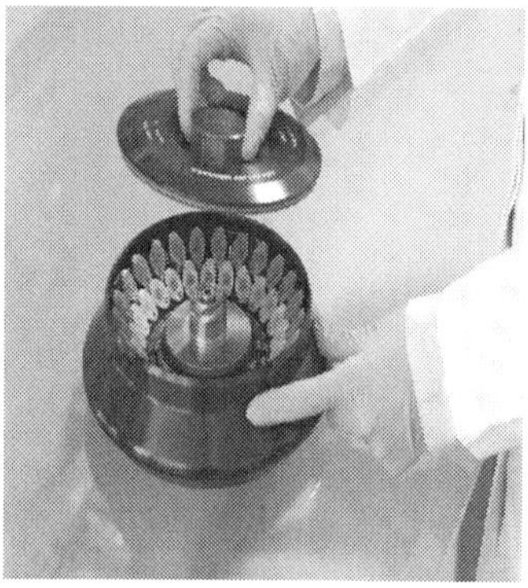

Centrifuges are used to spin samples in an attempt to separate or isolate solid elements from liquid elements of the samples such samples may include blood, cells or cell organelles.

Microcentrifuge tubes, commonly referred to as Eppendorf tubes, are the containers used to hold and store the samples. They are short tubes with a tapered end and attached cap on top.

TIPS-using Centrifuge

Inspect centrifuge bottles/tubes for cracks. Because a centrifuge can spin at such high speeds, a liquid sample can easily become an aerosol if it is not properly contained. Cracked tubes can fracture at high speeds or at the very least, leak sample into the rotor. Discard broken or cracked tubes to prevent them from accidentally being used in the future.

Cap tubes with the proper lid. Tubes that have been specified for use with a centrifuge have a proper cap or lid that seals the tube. Using a different lid or covering such as aluminum foil, sealing tape can easily lead to spills within the centrifuge. Making an aerosol of your sample can be hazardous to your health and the environment. In addition, it can lead to loss of precious samples.

Clearly label tubes for identification. You may know how you placed your tubes into the centrifuge before started spinning, but at the end of the spin you will not be able to tell them apart. Make sure you put some sort of label on each tube so you know which sample is which. If you have a balance, be sure you label it as such. It is best to label the tube directly instead of using a sticker. A sticker can fall off during the spin making identification difficult. Be sure that the marker is permanent and not easily rub off.

Placing a clean tube in the centrifuge. If you are working with a biohazardous it is a must to wipe the tube prior to placing it into the rotor. This is to limit any possible spills or aerosol formation from the sample into the rotor. In addition, it will also reduce the damage cause to the rotor.

Make a counterbalance. It is important to make a counterbalance. It is because at high speeds, a centrifuge can easily become unbalanced if equal masses are not placed opposite to each other in the rotor. Therefore, remember to balance the masses of the tubes and not the volumes. Always weigh the tube with your sample and record the mass, add water to a separate empty tube until the mass is the same as your sample. The important to counterbalance is because the unbalanced tubes can lead to permanent damage of the centrifuge and can be hazards if the rotor breaks free from the rest of the centrifuge. Balancing is extremely important at higher centrifugation speeds. Modern centrifuges will automatically turn off if it senses an unbalanced load, but older ones may not have this feature and it begins to shake or wobble. If this situation was noticed the user need to off the balance manually by pressing the stop button immediately.

Placing the centrifuge. It is important to place the centrifuge on a solid flat level due to its high speeds during the spins. The solid surface should be enough to support the weight of the centrifuge.

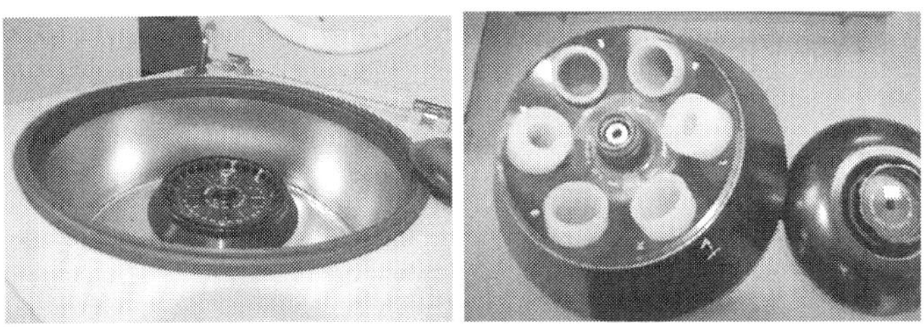

Choice of rotor. The rotor is the piece that hold and spins your sample. There are various types of rotor, some rotors have arms that you can attach smaller buckets, while other rotors are one piece that user can place the tubes directly into. Rotors are designated for specific tubes sizes and speeds. Suitable rotor need to be used to fit the experimental tubes and within the limit of speed needed.

Centrifugation speed. Centrifugation speed is often given in rotations per minute (rpm), how many times the rotor completes a full rotation in one minute, in relative centrifugal force (RCF) or in the G-force on the rotor. Modern day centrifuges are digital and have settings for both. The speed at which the user spins their samples is dependent upon what they are spinning according to their experimental protocol.

Safe Distance. Do not move the centrifuge during the use as it can cause an imbalance and lead to injury. Larger centrifuges can spin at very high speeds and are extremely dangerous if they become imbalanced. Do not stand near to the centrifuge while spinning.

Opening the lid. Many modern centrifuges have a lock on, that will not unlock until the rotor has come to a stop state. Older centrifuges may not have this feature so make sure the rotor is no longer moving before opening the lid. Keep the power cords away from the ground to avoid trip hazards that could move the centrifuge during spinning.

Removing the tubes. Remove the tubes gently so that the separated suspensions do not mix again. In addition, also check if there is any leakage cause by broken. If a spill has occurred, clean the rotor and buckets immediately.

MaintenanSce. To keep the centrifuge clean and running smoothly, always clean the rotor and wipe down the centrifuge at the end of each run. When you are finished, leave the lid open so it can air out and remain dry.

Incubators

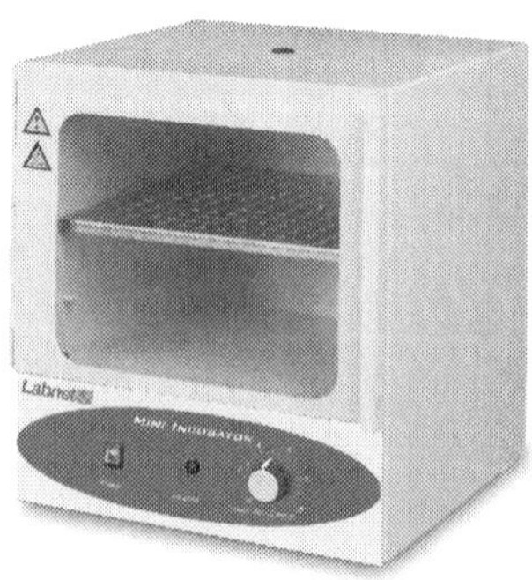

An incubator is an equipment used to grow and maintain microbiological cultures or cell cultures. The incubator maintains optimal temperature, humidity, carbon dioxide (CO_2) and oxygen content of the atmosphere inside the incubators. Incubators are important for a lot of experimental work in cell biology, microbiology and molecular biology. Various brands/types of incubators are commercially available. Certain incubators are specially made to cater certain industry such as poultry industry for hatching purposes. The simplest incubators are designed with insulated boxes with an adjustable heater. It typically going up to 60 to 65 °C (140 to 150 °F) and can go slightly higher up to no more than 100 °C. The most commonly used temperature both for bacteria for example *E. coli* as well as for mammalian cells is approximately 37 °C and another example the budding yeast Saccharomyces cerevisiae, a growth temperature of 30 °C. These organisms grow well under such conditions. There are incubators can also be set to lower temperature. This is important in the cultivation of mammalian cells which grows in lower temperature.

PIPETTE

A pipette is a common laboratory tool used in chemistry, biology and medicine to transfer a measured volume of liquid. There are several designs of pipettes that serve various purposes with differing levels of accuracy and precision. From a very basic single piece of glass pipettes to more sophisticated, adjustable, electronic or robotics pipettes are available commercially. Many pipettes operate by creating a partial vacuum above the liquid-holding chamber and draw up and dispense liquid by releasing this vacuum. Measurement accuracy varies greatly depending on the style. While, propipette is used to draw liquid from certain pipets. There are various types of propipette was described below.

Types of Propipette

Propitte

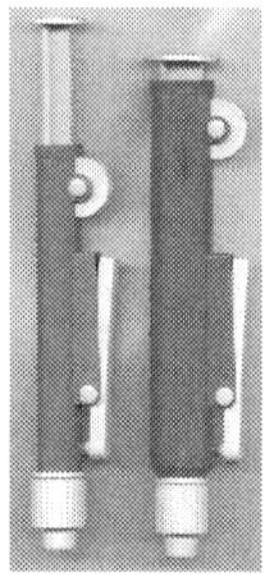

A manual propipette is used by adjusting the wheel with the thumb to draw liquid or any solvent into graduated pipets and volumetric pipets.

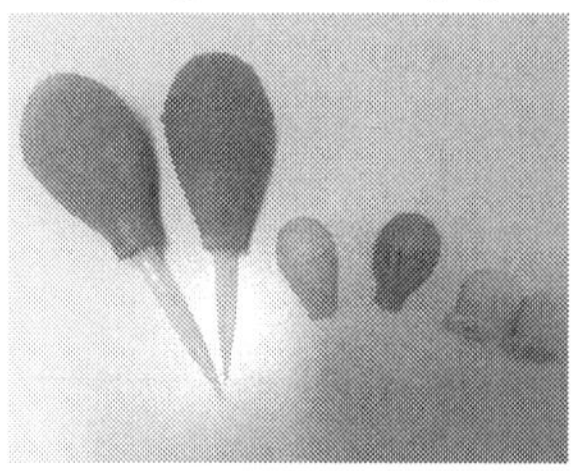

This type of manual propipette is adjusted by squeezing the bulb and it is also can be used with graduated pipets and volumetric pipets.

An automatic propipette was designed to ease the user. It can be adjusted by pressing the button and toggling the switch.

An automatic propipette is also another example of propitte and it can be adjusted by pulling and releasing the triggers.

Types of pipets

Volumetric pipettes

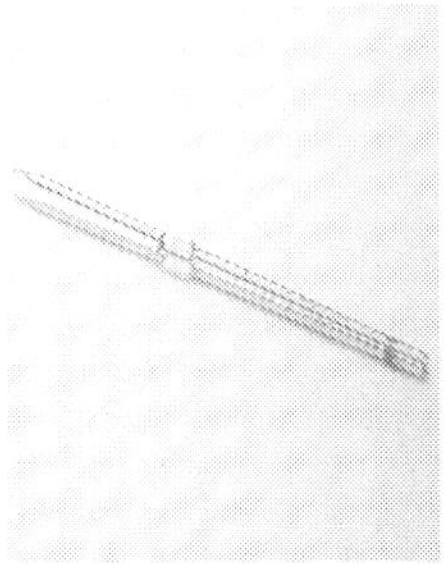

Volumetric pipettes or also known as bulb pipette was designed to allow the user to precisely measure a volume of solution. These pipettes comes with a large bulb with a long narrow portion above with a single graduation mark that is calibrated for a single volume. Typical volumes that are available are 10, 25 and 50 mL. It is used to make laboratory solutions from a base stock as well as prepare solutions for titration.

Graduated pipettes

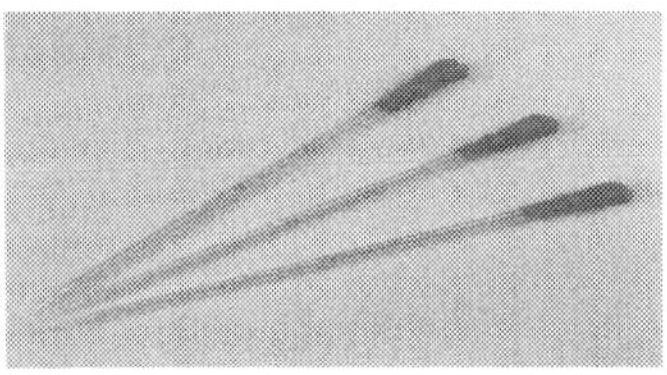

Graduated pipettes are a type of macro-pipette design in a long tube with a series of graduations with various calibrated volumes. A source of vacuum is required to draw liquid into the tube.

Pasteur pipette

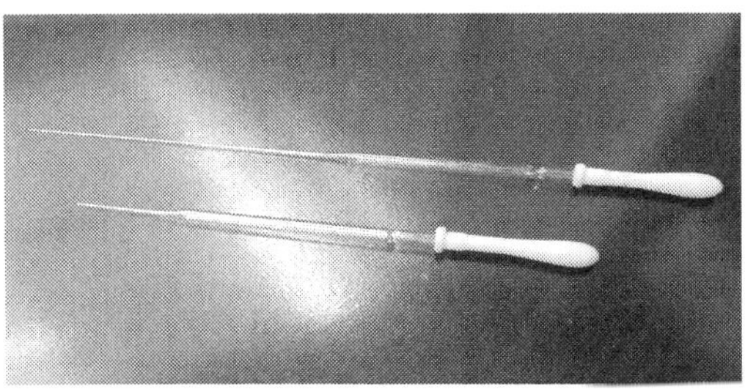

Pasteur pipettes can be found either in plastic or glass pipettes. It is attached with rubber bulbs that is separated from the pipette body. It is used to transfer small amounts of liquids. The proper handling of the pasteur pipettes is by holding the rubber bulb between the thumb and index finger and control the glass by holding it between the ring finger and the pinky.

Transfer pipette

Transfer pipettes, are similar to Pasteur pipettes but are made from a single piece of plastic and their bulb can serve as the liquid-holding chamber.

Glass micropipette

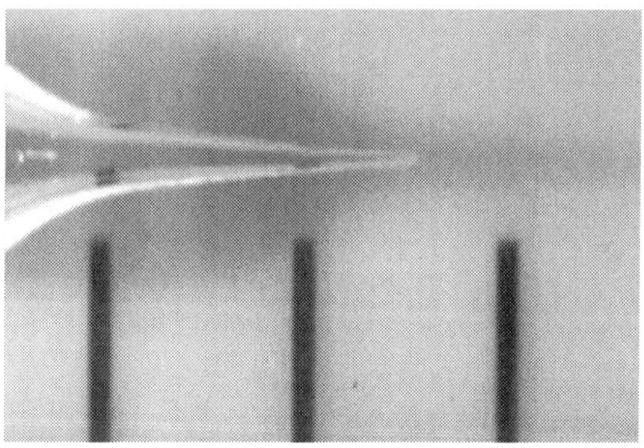

Borosilicate glass micropipette are used to physically interact with microscopic samples, during the procedures of microinjection and patch clamping. Most micropipettes are made of borosilicate, aluminosilicate or quartz with many types of sizes. Each of these compositions has unique properties and this property will determine the suitable applications. Is also can be fabricated in a micropipette puller and are typically used in a micromanipulator.

Positive displacement pipette

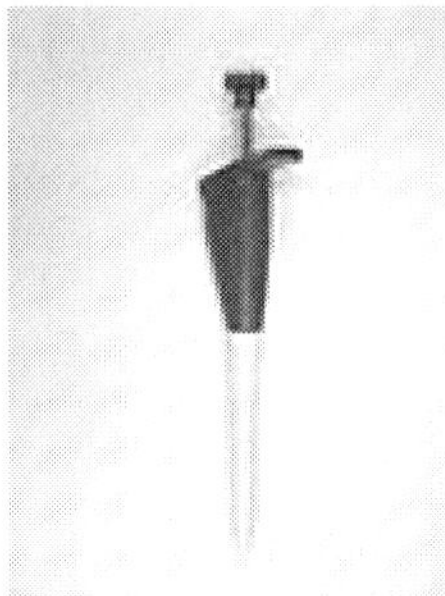

Positive displacement pipettes are mainly used for volatile or viscous substances at small volumes. It uses microsyringe (plastic) as disposable tip and a plunger which directly displaces the liquid.

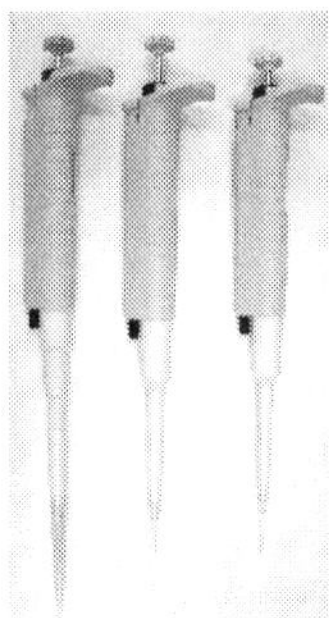

Adjustable pipets

Adjustable pipettes are designed in various range such as

a)20–200μl,

b)2–20μl,

c)100–1000μl and

d)25mL,

e)10mL,

f) 5mL and

g)2mL

h)1-5ml and

i)100–1000μl with locking system.

And each type of pipets comes with variety of tips. The four standard sizes of micropipettes correspond to four different disposable tip colors.

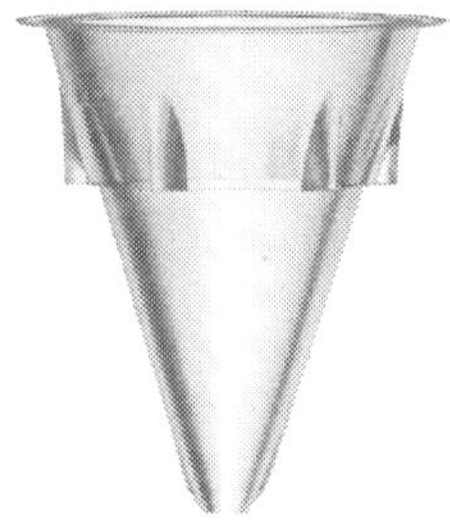

These pipettes are operated by piston-driven air displacement concept. A vacuum is generated by the vertical travel of a metal or ceramic piston within an airtight sleeve. As the piston moves upward, driven by the depression of the plunger, a vacuum is created in the space left vacant by the piston. The liquid around the tip moves into this vacuum (along with the air in the tip) and then be transported and released when necessary. These pipettes are capable of being very precise and accurate. However, since they rely on air displacement, they are subjected to inaccuracies caused by the changing environment, particularly temperature and user technique.

For these reasons, this equipment must be carefully maintained and calibrated and users must be trained to use it correctly. Checking the calibration at least every six months or quarterly is depends on the frequency of usage and the seriousness. It is the act of determining the accuracy of a measuring device by comparison with a standard. Pipette calibration is essential to ensure that the instrument is working according to expectations and as per the defined regimes or work protocols.

Specialized pipettes

Certain pipettes are specialized to cater certain needs. Mainly all are calibrated to ISO volumetric A grade standards. This arrangement provides improvements in precision, handling safety, reliability, economy and versatility.

Ostwald–Folin pipette

Is a special pipette used in measuring viscous fluid such as blood.

Ostwald-Folin pipette is common in medical technology laboratory setups together with other pipettes.

Microfluidic pipette

A microfluidic pipette is housed in a manifold holder. The pipettes are made from polydimethylsiloxane (PDMS) which is formed using reactive injection molding. Pneumatic actuation is used to keep all tubing free from contamination. The colored solutions highlight the solutions loaded into the wells of the PDMS pipette. A recent introduction into the micropipette field integrates the versatility of microfluidics into a freely positional pipette platform. A localized flow zone is created at the tip of the device, to allow a constant control of the nanoliter directly in front of the pipette. Interfacing of these pipettes using pneumatics enables multiple solutions to be loaded.

Microfludic

Microfluidics deals with fluids that are geometrically constrained to a small, typically sub-millimetre, scale. Typically, micro means one of the following features:

- Small volumes (μL, nL, pL, fL)
- Small size
- Low energy consumption
- Effects of the micro-domain

Typically fluids are moved, mixed, separated or otherwise processed by 2 active (micro) components such as micropumps or microvalves. Micropumps supply fluids in a continuous manner. Microvalves determine the flow direction or the mode of movement of pumped liquids. Often processes which are normally carried out in a lab are miniaturised on a single chip in order to enhance efficiency and mobility as well as reducing sample and reagent volumes. Advances in microfluidics technology are revolutionizing molecular biology procedures for enzymatic analysis, DNA analysis and proteomics. The basic idea of microfluidic biochips is to integrate assay operations such as detection, as well as sample pre-treatment and sample preparation on one chip. An emerging application area for biochips is clinical pathology, especially the immediate point-of-diagnosis of diseases. In addition, microfluidics-based devices, capable of continuous sampling and real-time testing for biochemical and dangerous pathogens analysis. Microfluidic technology has led to the creation of powerful tools for biologists to control the complete cellular environment.

Extremely low volume pipettes

The pipette is made of a carbon shell, within which is an alloy of gold-germanium. The pipette was used to learn about how crystallization takes place.

Robots

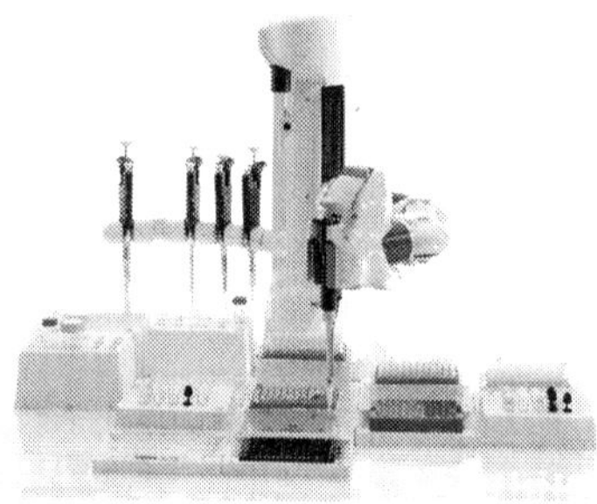

An example of pipettes manipulated by an anthropomorphic robot. Robotic pipette are capable of manipulating the pipettes as humans would do.

GROWING BACTERIA IN PETRI DISHES

Growing bacteria is also a biological sample processing once a bacteria sampling was carry out.

Simply take samples around you and see what bacteria will grow in an agar-filled Petri dish.

This activity will prove that mother was right...

"Wash your hands with soap and warm water!"

A Petri dish prepared with nutrient agar is an ideal way to reveal the bacteria presence all around you. You may not believe what you find entirely on all the spaces.

You'll need a clean, microwave-safe container to mix the agar with water and then boil it.

- ½-teaspoon agar (about 1.2 grams)
- ¼ cup (60 mL) of hot water

These proportions make enough nutrient agar to prepare two Petri dishes.

Stir these together and bring this mixture to a boil for three minutes to completely dissolve the agar.

CAUTION: If you are using a microwave oven to boil the mixture, be careful not to let it boil over. The mixture should be boiled until it is clear with no particles floating in it after boiling.

Remove the mixture from the microwave and allow it to cool for 3 to 5 minutes before moving on to the next step.

Take the lid off of the Petri dish (the lid is larger than the dish) and carefully cover the bottom-half of the Petri dish with warm nutrient agar mixture.

Loosely cover the bottom portion (set the lid slanting to ensure escape of excess moisture) and allow the mixture to cool and harden for at least an hour. Moisture coming from the agar can be a problem.

NOTE: Agar needs to boil for a certain amount of time as it is just like a gelatin. In case, there is some gelatin which is still floating in it re-boiled the whole mixture once more. Keep an eye during this procedure to avoid boiling over until no "floaties" in it. Pour the mild-hot agar mixture back into the dishes as you did before and it should solidify within an hour. For bacterial collection, one can use wet sterilized cotton swab. User can collect some bacteria on the end of a cotton swab. This can be done by swabbing the end of cotton in your mouth and then to lightly draw a zig-zag with it on the gelled agar. Replace the lid and label the dish with the date and the name of the item you tested.

You can also try on something even more challenging space such as on a computer, a cell phone case, the pump handle of a soap dispenser, or the TV remote control. Try to cover the entire cotton end of the swab. You will see less bacteria in case the space was recently cleaned and sterilized. Incubate the petri dish upside down to prevent moisture from dripping onto the growing colonies.

Remember: Do NOT open the dishes once things begin to grow.

For disinfecting the petri dish after the studies, seal all the Petri dishes into larger zipper-lock plastic bags and dispose the entire sealed bag in the trash after autoclaving it.

CHAPTER

5 TOOLS AND TECHNIQUES IN BIOLOGICAL SAMPLE ANALYSIS

INTRODUCTION

There are two types of biological sample analysis. One of it is external analysis of the sample and the other one is internal analysis. Both analysis used different types of tools. A complete description of the tool selection for external and internal analysis will be described in this chapter.

EXTERNAL ANALYSIS

Binoculars

In case you need to work in the wildlife, the study of animals takes place in their natural habitat. To observe them in the wild without any disturbance, binoculars are used to keep a safe distance from animals that could be dangerous.

Magnifying Glass

To closely observe small creatures and insects with naked eyes is not easy, which is why a magnifying glass is one of the essential tools. Magnifying glasses are used to study insects.

INTERNAL ANALYSIS

Microscopes

Light Microscope

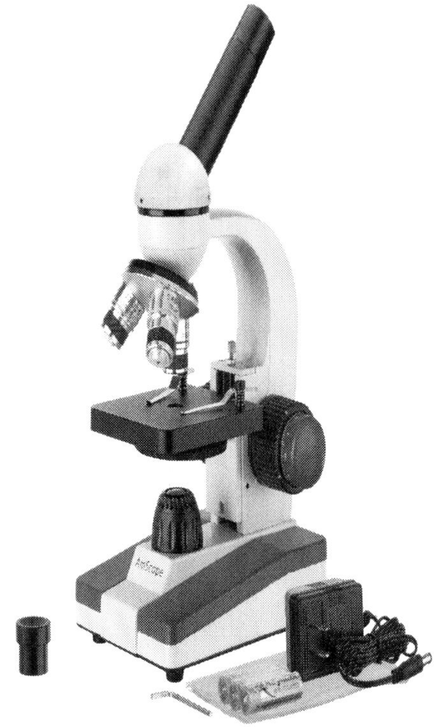

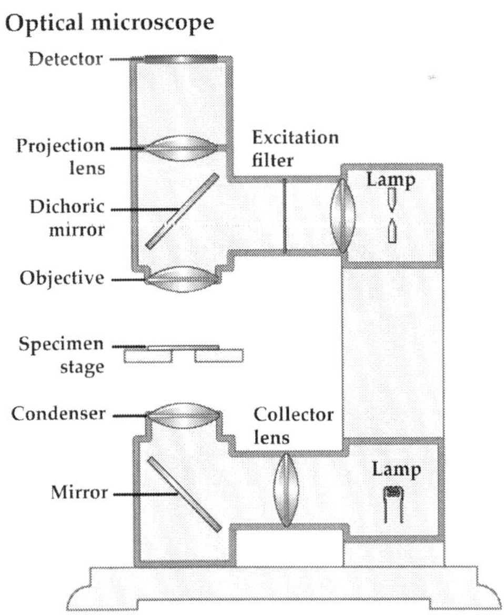

Figure was adapted from Molecular Cell Biology, Sixth Edition © 2008 W.H. Freeman and Company

Light microscopes, also known as optical microscopes, are one of the most basic tools biologists use. Light microscopes use several lenses and visible light to magnify small specimens being studied in a biological lab such as bacteria, plant and animal tissues. Samples are placed on microscope slides with some form of liquid for preserving the specimen and for better view. A complete operating guide was describe in the next part.

Microscope: Operating Guide

General Use:

Always placed the microscope on a flat surface. For beginners, the user can learn to focus the microscope mouse over the picture (X) that was marked by marker, crayon or a sharpie marker.

- Firstly, user need to turn the revolving nose piece to engage the 10X objective lens. User need to make sure that the revolving nose piece stops with a click.
- User must gently place a prepared slide on the stage in the specimen holder clips by lowering the stage using the coarse adjustment. User need to ensure that the specimen holder clips placed on the slides gently as it may break the slide if released in harsh.
- To move the mark on the slide into the light path user need to turn the X-axis and Y-axis adjustment knobs. User should not move the stage manually without the adjustment knobs and the knobs should not be forced as it has the limitations.
- Users can adjust the brightness with the light intensity knob with switching the main switch.
- User shall turn the coarse adjustment knob to bring the specimen into focus by looking through the right eyepiece with user`s right eye. Once user have optimized the focus with the coarse adjustment controls, they can use the fine adjustment knob to improve the focus.
- By looking through the left eyepiece with user`s left eye, they can turn the diopter adjustment ring to focus the specimen.
- By using the eyepieces users can adjust the interpapillary

distance. By looking through the eyepieces users can manually adjust the space between the eyepieces until the left and right fields of view coincide completely. This interpapillary adjustment is specific to each user.

- Users shall manipulate the field iris diaphragm to be centered by turning the centering screws on the compensating lens.
- Users shall engage the pre-focusing lever by rotating it. The lever ensures that the objective does not come in contact with the specimen and to simplify the focusing. To be safe, users set the lever to an upper limit by coarse adjustment. Then one can start focusing with the fine adjustment knob. This will ensure that the lens is not hit by the lever. The microscope is designed to allow the microscope user to change the objective lens to view the image with little or without fine focusing adjustments. Some microscope like par focal microscopes allow the microscope user to change the objective and view the specimen with minimal focusing, using only the fine adjustment knob.
- User can engage the objective that need to be used for observation by turning the revolving nosepiece. User can refocus the rotating coarse adjustment knob to reach the pre-focusing position, with a limitation of the pre-focusing lever. Try make fine adjustments with the fine adjustment knob. Safer tips are to always move the objective away from the slide.
- Users also need to adjust the light intensity to the desired level, and then readjust the fine focus.
- User should also adjust the field iris diaphragm.
- It is a must to adjust the aperture iris diaphragm. It is because the iris aperture diaphragm controls the cone of light that reaches the objectives and therefore controls the numerical aperture. Matching the numerical aperture of the illumination system (given on the iris diaphragm) with that of the objective will provides a better image resolution and contrast, and also increases the depth of focus.

Usage of Oil Immersion

The presence of oil between the objective and the specimen will decrease the diffraction of light that passes through the objective. This is important for resolution of objects at 1000X. There are few steps need to be followed in order to use the objective lens with oil immersion:

Put in bullet point each sentences.

- Place a drop of immersion oil on the center of the slide where the specimen to be viewed is placed.
- Then turn the revolving nosepiece to engage the oil immersion objective.
- With using the fine adjustment control try focus the image.
- Make sure the oil on the specimen must be free from bubbles to avoid impaired image.
- To remove bubbles, slightly move the oil immersion lens back and forth over the specimen.
- Once done, it is very important that all oil be cleaned off from the lens before storing the microscope.

Phase Contrast Microscope Operating Guide

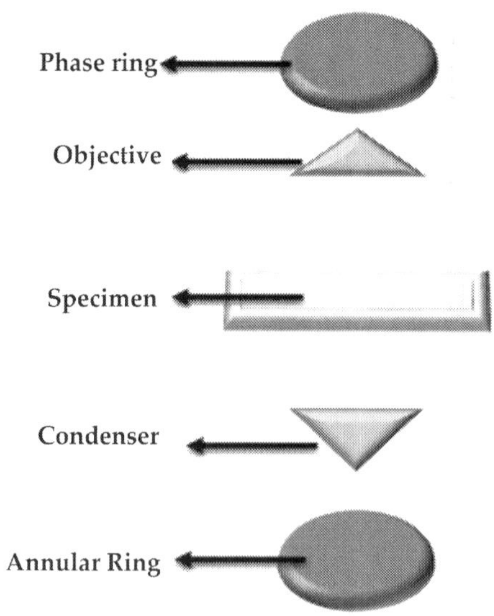

Phase contrast microscopy allows better viewing of unstained specimens, making use of small differences in the refractive index of the specimen. In addition to being "bent," light passing through a substance is slowed so that it is out of phase with light that did not pass through the specimen. The difference between the diffracted, or slowed, light and the direct light is difficult to detect in a microscope without phase contrast objectives and a phase contrast sub-stage condenser.

In a phase contract microscope, the difference between the direct and diffracted light is increased. A ring in the sub-stage condenser of the phase contrast microscope focuses a hollow cone of light on the specimen. Undiffracted light, or direct light, in this cone speeds up as it passes through a phase-shifting element in the objective lens. Light passing through the specimen (and diffracted) is combined with direct light that passed through the phase-shifting element. The result of combining the two beams of light, is a darker or brighter spot in the image, depending on the difference in the phases of the two light beams.

Annual Ring Laboratory Instrument: Microscope

Microorganisms are too small to viewed. In order to view them, we need to magnify an image or collect a large number of the microbes in one place. We can observe a large number of organisms when we look at colonies on a plate. We look at individual organisms using a microscope. Standard microscopes amplify the image through two lenses. One lens is in the eyepiece and the other is in the objective. To calculate the overall magnification, one multiplies the magnifying power of each lens together. Eyepieces usually contain lenses that amplify ten times. Objectives vary in a magnification and most microscopes contain several objectives from which the microscope user may choose.

(10X eyepiece lens)(40X objective lens) = 400X total amplification

The top magnification for a light microscope is 1000X. Working distance is the space between the objective and the specimen or slide. The focal distance increases with the working distance. The magnification is inversely related to the working distance. Therefore, the 100X objective would have the smallest working distance.

The top magnification for a light microscope is about 1000X. Above this, the limits of the optical system produce a fuzzy image. This is related to the resolving power, or resolution, of the microscope. The resolving power or resolution of the microscope is the ability of the microscope or optical instrument to distinguish between two very close objects.

Resolving power = wavelength of light used/(Numerical aperture of the objective + Numerical aperture of the sub-stage condenser).

The numerical aperture is a measure of the size of the cone of light that may be gathered by the objective. The largest cone of light is needed for the highest magnification. With low magnification, the working distance is increased, less definition is required and the numerical aperture setting is lower. The microscope user can control the light as it passes through the sub-stage condenser and may use this to change the numerical aperture.

Electron Microscope

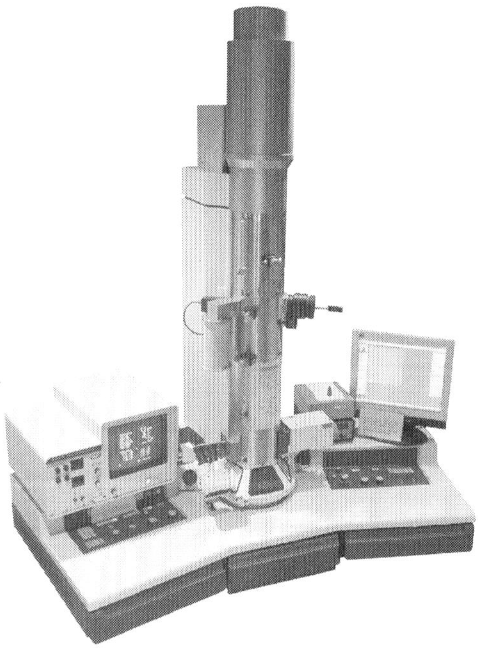

Electron microscope is a sophisticated form of microscope, which can provide a higher magnification of the object up to 100,000 times or even more. The electron microscope employs an electron beam originate from a cathode, instead of light as in the compound microscope. It uses coils instead of lenses.

Electron Microscope

There are two types of electron microscopes used in biological studies:

Transmission Electron Microscope (TEM)

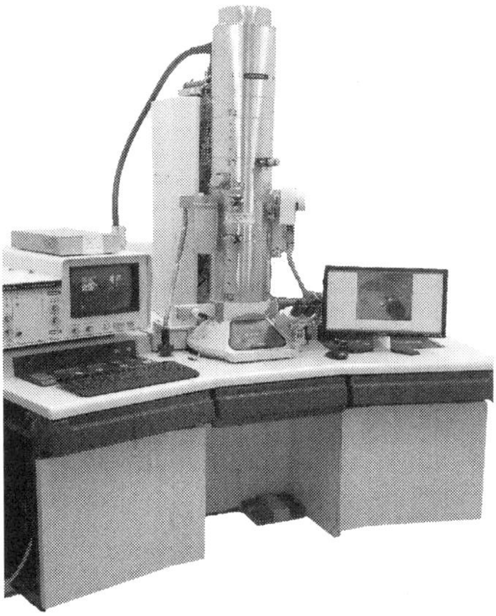

Scanning Electron Microscope (SEM)

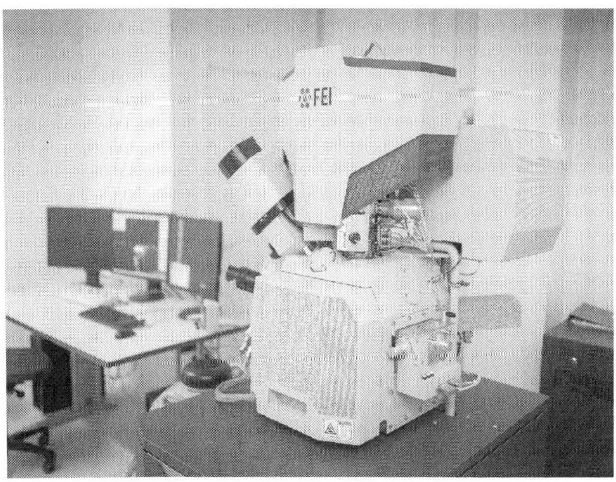

It used for the study of three-dimensional surface view of a given specimen. Its resolving power is around 2 to 10^0A.

Cellular imaging system

There are new generation of binocular laboratory microscope in modern ergonomic design and latest LED technology. They are developed for use in laboratories, doctor offices and remote areas for education and research.

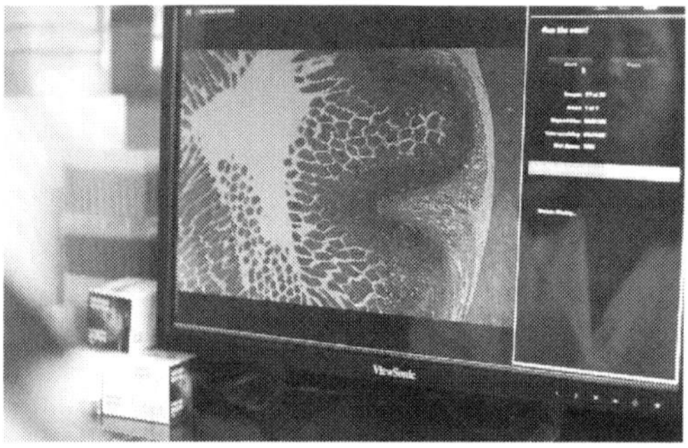

Chromatography

Chromatography is a technique for separating mixture of components by partitioning between two solvent system which is stationary phase and liquid phase.

The below figure shows 5 types of chromatography divided based on the nature of the solid phase:

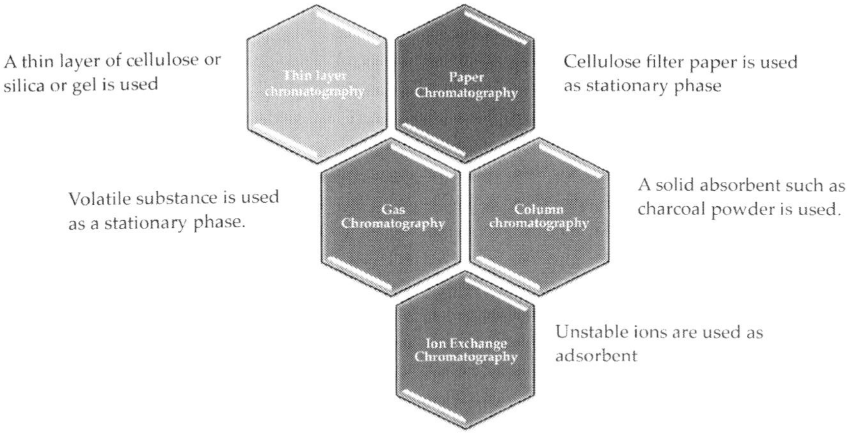

Colorimetry

Colorimetry is a technique to measure the density of a colored substance in a solution. The instrument itself called colorimeter. Monochromatic filters are used and lights rays of specific bands of wavelength are passed through the solutions. Any compounds that present in the solutions will reduces the intensity of the transmitted beam due to the absorption and this value will be recorded by the photocell.

Freeze Etching

It is a technique in which cells and tissues that are frozen were cut with a microtome knife. This technique helps in studying cell surfaces and intercellular organelles.

Tools for DNA and RNA analysis

Gel Electrophoresis

Gel electrophoresis is a laboratory method that used to separate mixtures of DNA, which is also used for RNA, or proteins separation according to molecular size. In gel electrophoresis, the molecules to be separated are pushed by an electrical field through a gel that contains small pores. The molecules travel through the pores in the gel at a speed that is inversely related to their lengths. This means that a small DNA molecule will travel a greater distance through the gel than a larger DNA molecule. One end of the gel electrophoresis has a positive charge and the other have a negative charge. Since DNA and RNA are negatively charged molecules, they will be pulled toward the positively charged end of the gel. Finally, after the DNA and RNA, or protein molecules have been separated using gel electrophoresis, bands representing molecules of different sizes can be detected through staining.

Spectrophotometry

The nature of the interaction between light and matter depends on the wavelength of the light and the structure of the matter. Useful wavelengths for spectrophotometry range from 185 to 3,000 nm. Absorption of longer, low energy wavelengths (infrared) stretches and bonds in molecules. The resulting complex patterns of absorption at different wavelengths are used in IR spectroscopy to identify chemical structures. Shorter wavelengths (X-rays and gamma rays) are sufficiently energetic to strip electrons from molecules. In the visible/UV range used in spectrophotometry, molecules absorb light energy with the movement of electrons to a higher orbital. Note that molecules absorb only those wavelengths of light that contain just the right amount of energy to move them to a higher energy state, resulting in peaks and valleys when absorbance is plotted against wavelength. The amount of light passing through the sample decreases exponentially (by percentages) as the concentration increases.

Spectrophotometers present the results either as percent transmittance or as absorbance. Percent transmittance is simply the percentage of incident light reaching the photocell. The second scale on the spectrophotometer reads are indicate as absorbance (A).

1. Be sure to check the cell holder is empty.
2. Turn on the spectrophotometer using the power switch of the instrument.
3. Allow the spectrophotometer to warm up for 30 minutes.
4. Press A/T/C to select the absorbance or % transmittance mode.
5. Set the wavelength to the desired value. Holding the key down will cause the wavelength to change more quickly.
6. Insert the blank into the cell holder with the clear walls of the cuvette to the front and back of the instrument.

Close the sample compartment. The blank is used to zero absorbance readings due to imperfections in the cuvette and for the absorbance of light by the medium or carrier solution. The procedure you are using will contain specific instructions about the composition of the blank.

Cuvettes

- Press to blank the instrument.
- Remove the blank from the cell holder, insert the sample and close the sample compartment.
- Record the instrument reading.
- Remove the sample from the sample compartment.

Real Time PCR

Consumables and Equipment

Pipettes, sterilized tips, microcentrifuge, racks, fine tipped markers; MicroAmp® Optical 8-Tube Strip (0.2 ml)

Reagents

cDNA synthesis kit, Primers, Real-Time buffers

Instruction: cDNA Amplification

- Thaw RNA in ice while cDNA synthesis kit at room temperature.
- Mix RNA and primers that comes in the kit.

- Incubate at 65°C

Instruction: Assembly of reactions

- Thaw primers, reagent and buffers at RT and cDNA samples in ice.
- Design a real-time RT-PCR assay using MicroAmp® Optical 8-Tube Strip (0.2 ml). We recommend a triplicate (3 wells) for each sample.
- Carefully calculate the amount of volume of primers, reagents, buffers and nuclease free water needed. Assemble a reaction mixture for each gene of interest in a microcentrifuge tube.
- Accurately aliquot the same volume of the reaction cocktail to each well accordingly.
- Spin down the microcentrifuge prior placing it the system.

Instruction: Programming and performance of ABI 7500

Be sure to read the manual book that comes with the equipment prior to handling any equipment.

ELISA

Elisa or enzyme-linked immunosorbent assay is an analysis that uses antibodies and color changes to identify targeted substance. It has been used as a diagnostic tool in medicine, plant pathology and also in a quality-control check in various industries.

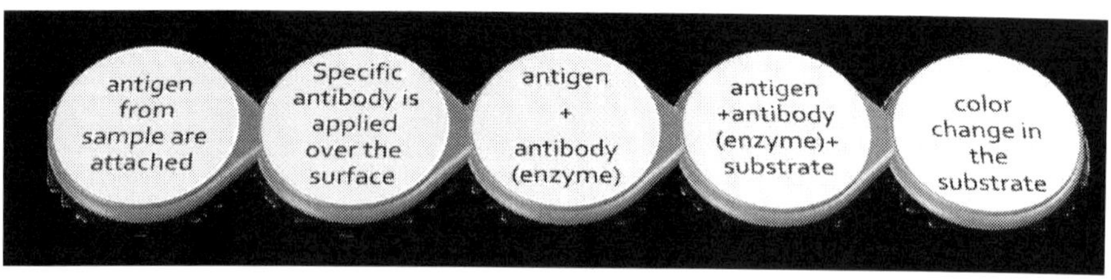

One antibody with specificity for a particular antigen is required. The sample with an unknown amount of antigen is immobilized on a solid support in the microtiter plate either non-specifically by adsorption or specifically by capturing on another antibody specific to the same antigen in a form of sandwich-like ELISA. Once the antigen is immobilized, the detection antibody is added to form a complex with the antigen. Through bioconjugation, the detection antibody covalently linked to an enzyme. Between each step, the plate is typically washed with a mild detergent solution to remove any proteins or antibodies that are non-specifically bound. After the final washing step, the plate is developed by adding an enzymatic substrate to produce a visible signal, which indicates the quantity of antigen in the sample.

ELISA can perform other forms of ligand binding assays instead of strictly immuno-assays, though the name carried the original immuno because of the common use and history of the development of this method.

TIPS AND TECHNIQUES- Elisa

- Essentially requires ligating reagent that can be immobilized on the solid phase.
- A detection reagent that will bind specifically.
- Use an enzyme to generate a signal that can be properly quantified.
- The ligand and its specific binding counterparts remain specifically bound by antigen-antibody interactions to the solid phase.
- The nonspecific or unbound components are washed away.
- ELISA plates are not easily reusable as the reaction products immunosorbent on the solid phase will remained.

ELISA as a heterogeneous assay, first will absorb the liquid sample onto a stationary solid phase. A special binding properties between the liquid sample and the stationary phase is initiated by an addition of multiple liquid reagents followed by incubation and washing. This is to enable ELISA to separates certain component in the analytical reaction mixture. In the final stage, there will be some optical changes that will be used to measure and quantify the targeted analyte. The qualitative reading usually based on the detection of the intensity of the transmitted light by spectrophotometry. The sensitivity of detection depends on the signal during the analytic reactions. Since enzyme reactions are very well known by the 'lock and key' mechanism, the signal which generated by enzymes are linked to the detection reagents in a fixed proportions to allow accurate quantification – thus it is called "enzyme linked" reaction.

The analyte is also known as the ligand because it will specifically bind or ligate to a detection reagent. The ligand-specific binding reagent is immobilized and constructed as a multiple-well plate (ELISA plate). Conventionally, like other forms of immunoassays, the specificity of antigen-antibody type reaction is used, because it is easy to raise an antibody specifically against an antigen in bulk as a reagent. Alternatively, if the analyte itself is an antibody, its target antigen can be used as the binding reagent. When the presence of an antigen is analyzed, the name direct ELISA refers to a labelled primary antibody used and the term indirect ELISA refers to the antigen that is bound by the primary antibody which then is detected by a labeled secondary antibody. In the latter case, a sandwich ELISA is clearly distinct from an indirect ELISA by the secondary antibody.

ELISA may be run in a qualitative or quantitative methods. Qualitative results provide either positive or negative result for a particular sample. Two or three times the standard deviation is often used to distinguish positive samples from negative samples. In quantitative ELISA, the optical density (OD) of the sample is compared to a standard curve, which was generated by a serial dilution of a known-concentration solution of the targeted molecule.

Types of Elisa:

Direct ELISA

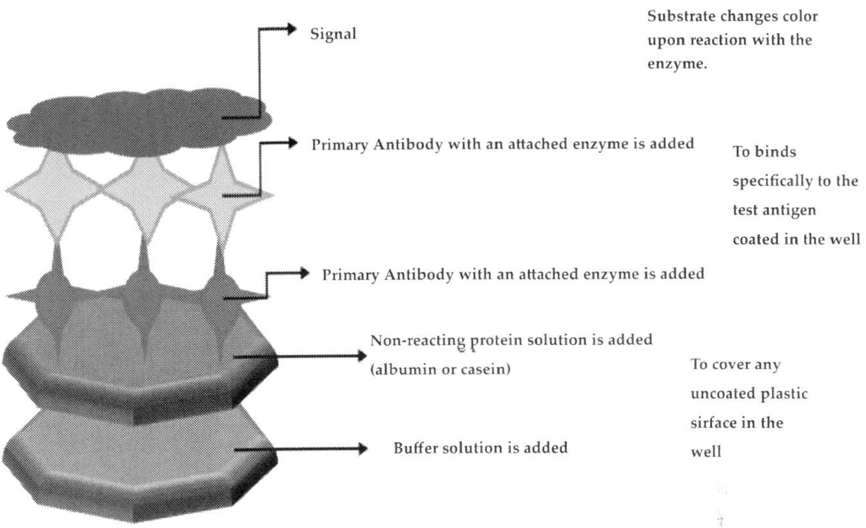

Concentration of the primary antibody present in the serum, the stronger the color change.

Disadvantage - Direct Elisa

The method of antigen immobilization is actually a non-specific format. When a serum is used as the source of test antigen, all proteins in the sample may stick to the microtiter plate well, so small concentrations of analyte in serum must compete with other serum proteins when binding to the well surface.

Sandwich Elisa (Indirect Elisa)

In sandwich or indirect Elisa the use of a secondary-antibody conjugate avoids the expensive process of creating enzyme-linked antibodies for every single antigen. By using an enzyme-linked antibody that binds the region of other antibodies, this same enzyme-linked antibody can be used in a variety of situations. Without the first layer of "capture" antibody, any proteins in the sample may competitively adsorb to the plate surface, lowering the quantity of antigen immobilized. Use of the purified specific antibody to attach the antigen to the plastic eliminates a need to purify the antigen from complicated mixtures before the measurement, simplifying the assay and increasing the specificity and the sensitivity of the assay. A sandwich ELISA used for research often need validation because of the risk of false positive results. The sandwich or indirect ELISA provides a solution to this problem, by using a "capture" antibody specific for the test antigen to pull it out of the serum's molecular mixture. A detectable signal such as fluorescent or electrochemical is measured to determine the presence and quantity of antigen.

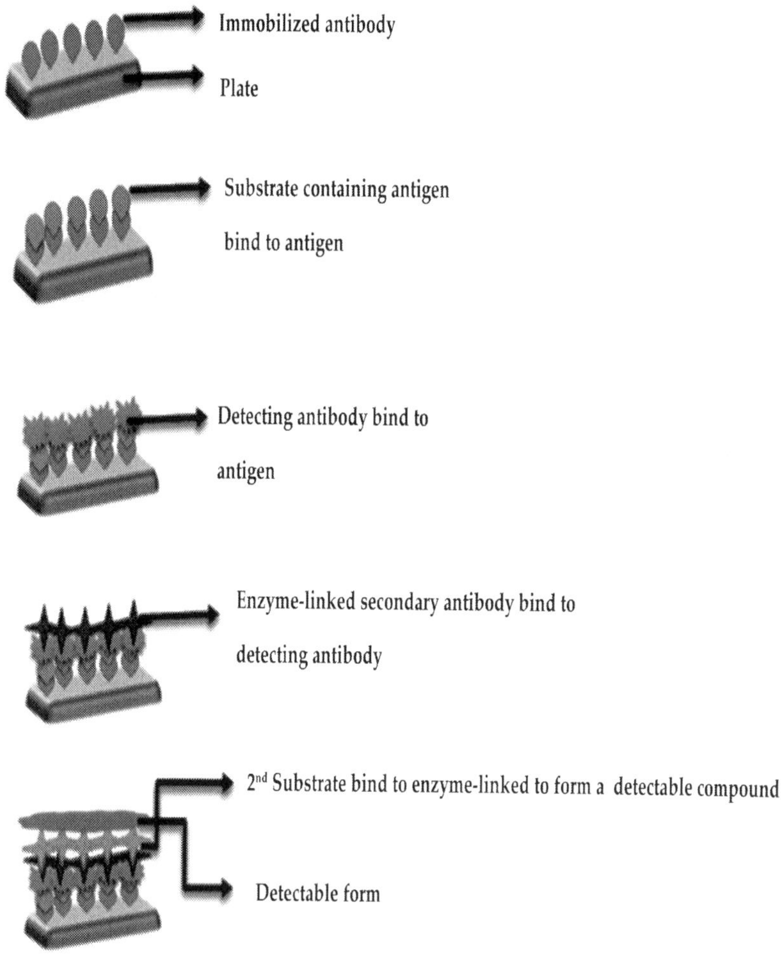

Competitive ELISA

The ongoing reaction is terminated to prevent eventual saturation of the signal. Some competitive ELISA kits that are commercially available includes enzyme-linked antigen rather than enzyme-linked antibody. The labeled antigen will compete for primary antibody binding sites with the unlabeled antigen sample. The less antigen in the sample, the more labeled antigen is retained in the well and it will lead to a stronger signal.

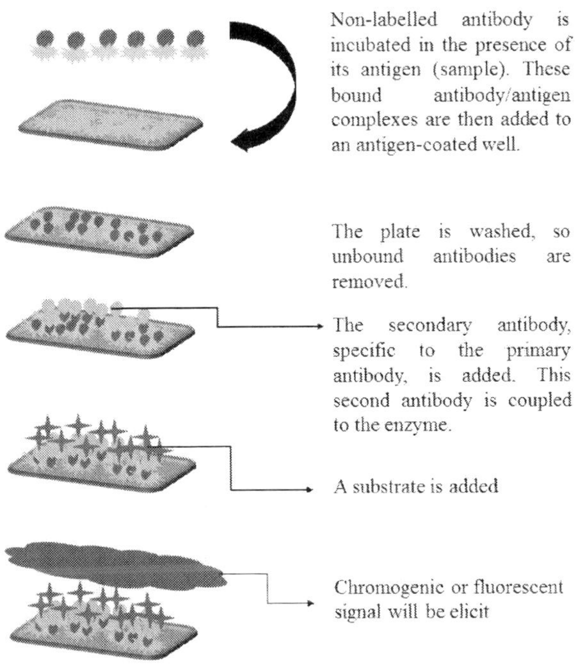

Polyacrylamide gel electrophoresis

Polyacrylamide gel electrophoresis (PAGE) is one of the most widely used laboratory methods to separate biological macromolecules such as proteins and nucleic acids. Macromolecules will be differentiated according to their electrophoresis mobility which is a function of the length, conformation and charge of the molecule. In general, macromolecules may be run in their native state or in denatured forms. To separate molecules based on their lengths, samples are run in denaturing conditions.

Sodium dodecyl sulfate

Sodium dodecyl sulfate (SDS) is used to negatively charge and to linearize the proteins. The binding between SDS and the polypeptide chain produce an even distribution of charge per unit mass. As a result, negatively charged proteins will migrate towards the positive electrode and will be separated by approximate size during electrophoresis. This procedure is called SDS-PAGE.

SDS-PAGE can be used to estimate the purity of protein samples. SDS is an anionic detergent that able to denature the proteins. The negative charges on SDS destroy most of the secondary and tertiary structure of proteins. Since the charge-to-mass ratio is nearly the same among SDS-denatured proteins, the final separation of proteins is almost entirely dependent on the differences in relative M_w of polypeptides. In PAGE, the relative migration distance of a protein (R_f) is negatively proportional to the log of its M_w. To be able to estimate the M_w of proteins on the SDS-page, proteins of known M_w need to be run simultaneously on the gel. These proteins are called Protein Standards. Electrophoresed SDS-PAGE will be further utilized into Western blotting. It is the combination of SDS-PAGE and antibody based detection. It is commonly used to detect proteins from complex biological mixtures through antibody application.

Protocol:

- User need to decide which percentage of gel that needed to separate the proteins.

For example:

- 4-8% gels are preferably used to separate proteins 100 to 500 kDa in size.
- 4-20% gels are preferably used to separate proteins 10 to 200

kDa in size.

- User must place the gel in a clean plastic electrophoresis chamber and corresponding gel holder.
- User need to prepare 1X SDS-PAGE Running Buffer.
- Fill the inner portion between the gel(s) and the gel holder with the appropriate 1X Running Buffer. Pour the remaining 1X Running Buffer into the outer chamber.
- For Sample Preparation:

If using a pre-prepared lysate, user need to thaw lysate and transfer 25 μL of lysate to a clean pre-labeled micro centrifuge tube. Add β-mercaptoethanol to a final concentration of 0.55M, for example 1 μL stock of β-mercaptoethanol was added in 25 μL lysate. It was mixed well by pipetting. All micro centrifuge tube must be labelled with sample description, volume and concentration of lysate.

For protein samples, user need to transfer it into a clean pre-labeled micro centrifuge tube and mixed with an equal volume of 2X sample buffer with 0.55M of β-mercaptoethanol. Concentration of protein sample should be sufficiently high in order to obtained a significant detection.

In preparing **molecular weight standards** for SDS-PAGE, user either use an unstained molecular weight standards or pre-stained molecular weight marker. In case the SDS-PAGE will be used into western blotting, a pre-stained molecular weight markers are needed.

It is important user remember to record lane, sample description, sample concentration, loading volume, loading amount and addition of reducing agent for all samples.

- Prior loading sample into the SDS-PAGE well, user need to place all micro centrifuge tubes containing samples into a heating block (set to 95°C) or water bath for 5 min.
- After heating, the aliquots need to be centrifuge for 3 min using a micro centrifuge to pellet any debris.
- Now user can load all samples into gel lanes starting with the molecular weight standards. Sample loading volumes should be from 5 μL to 35 μL per lane (depending on gel). If a protein concentration is from 100 μg/mL to 500 μg/mL, then sample amounts will range from 0.5 μg to 17.5 μg per lane. Generally,

1.0 μg is sufficient to visualize purified proteins and 10 μg is sufficient to visualize proteins in lysates on a Coomassie stained gel.

- Once all the sample was loaded into the gel well, cover the chamber and firmly connect both the anode and the cathode. The voltage was set on the electrophoresis power supply to a constant voltage of 150 V.

CAUTION! the functioning electrophoresis unit should not be touched while power is on. If user identified and leakage of the buffer from the unit, remember to TURN POWER OFF the power supply before making contact with buffer. Use care at all times.

- Try to allow the gel to electrophorese for 45-90 min. Once the dye front migrates out from the bottom of the gel, turn OFF the power immediately. Disconnect the electrodes and remove the cover followed by removing the gel holder from the electrophoresis chamber. User can now carefully remove the gel from holder so that breakage of the gel does not occur. Remove the gel from its plates and proceed with desired detection method.

TIPS AND TECHNIQUES- SDS

It is used to separate mixtures of proteins of variable complexity and allows you to determine sample protein composition, purity, whether you have a target of interest in your sample and even some of its structural characteristics. Gel electrophoresis is a basic tool by itself, but also is the stepping stone to other important techniques in the lab that going hand in hand with western blotting (WB). You can't have a good western blot without first having a good SDS gel. In this particular tip segment, we are discussing protein gel electrophoresis using precast gels and the parameters involved. Below are some pointers for optimizing your SDS-

PAGE results.

Safety First

- Typical voltages and currents used in electrophoresis are dangerous and possibly lethal.
- Ensure all cables and connections are routinely inspected for cracks and other defects from normal wear and tear that result in exposed wires and replace immediately.
- Avoid inserting or removing high-voltage leads with both hands to prevent shunt electricity through the body and make sure to always handle those when the power supply is turned off.
- When disconnecting the system, first wait until the voltage and current meters reach zero, turn off the power supply and then disconnect the gel box.

Choosing the Right Gel

Choosing the correct percentage gel is crucial. The larger the protein the lower percentage gel is preferable. Commonly used are 4-20% gradient gels that can cover a vast range of molecular weight sizes. Proteins ≥ 200 kDa will resolve better in 4-8% gels.

How Much to Load

Consider the concentration of your purified protein, lysate, or culture sample. Ideally, it is best to load ≤ 2 µg per well of a purified protein or ≤ 20 µg of a complex mixture like whole cell lysates if you are doing Coomassie stain only. Protein loading can be adjusted accordingly for more sensitive stains like silver and fluorescent staining or when doing western blot where you can do lower amounts. Loading too little sample will result in an incomplete picture of your target such as any contaminants or inaccurate detection of your target protein. Overloading can result in smears or streaks due to protein precipitating or aggregating. In general, improper loading will lead to inconclusive results.

Timing

After samples are loaded, you want to begin to run your gel immediately.

Delaying will result samples diffusing out of the wells. If you need to leave for a short amount of time, you can run your gel at low voltage to keep your samples from diffusing until your return. Now as for actually running your gel, follow manufacturer's recommendations of gel apparatus and pre-cast gel instructions. Typical conditions include runs at 100-150 volts for 40-60 minutes or until the dye front has reached the bottom of the gel. Letting it run too long will result in losing your lower molecular weight bands. Running too short and you will have poor resolution, especially in the low molecular weight range.

Heating Samples During Denaturation

Although not all samples need heating, it is critical for certain preparations like those containing membrane proteins. Thoroughly heating your sample not only completes the denaturing process but also ensures the dissociation of hydrophobic interactions, like those involving lipids. Not heating long enough results in incomplete denaturing, heating too long can cause aggregation. In general, heating at 95°C for 5 minutes should suffice. Following the denaturation step, spin down of samples at max speed for 2-3 minutes are critical to separate any particulates or aggregates that would interfere with gel loading and electrophoretic separation.

To Reduce or Non-Reduce

When considering protein molecules, which are built up of complex quaternary, tertiary, secondary and primary structures, we want to denature our samples to its relative simplest structure for electrophoresis. While SDS helps to break these structures, it also confers a negative charge to the amino acids giving a constant charge to mass ratio to all proteins and allowing separation to occur based on size. Nevertheless, SDS works well for secondary/tertiary structures but, for more complex structures like quaternary, we need to apply DTT (dithiothreitol) or BME (2-mercaptoethanol). These reducing reagents break disulfide bonds which are covalent and do not normally break using SDS alone. Both reducing agents work well, advantage of DTT is, it has less of an odor as BME, but breaks down faster. You can make your sample buffer with BME and have it freeze thaw over and over due to its stable properties. If you are looking to analyze molecular weight complexes then you would be performing a non-reducing electrophoresis and would not use reducing agents.

Proteins Properties

One must consider the protein's properties when preparing the sample for SDS-PAGE. Typically, you add your sample of known concentration to some amount of SDS sample buffer which is typically made up of SDS, glycerol, Tris-HCI pH 6.8, EDTA with or without reducing agent. In general, a 2X buffer stock can be used unless you have a diluted sample, in this case it would be beneficial to use a more concentrated stock (i.e., 5X or 6X) that allows you to load more samples in the same final volume. Proteins can also be further concentrated using trichloroacetic acid (TCA), acetone or methanol at the expense of some net loss. Depending on the gel's well size, you want to keep volume in mind, leading up to our next tips.

Gel Loading

When loading your gel, it may be easier to use your standard tips that you use on a daily basis. However, by using gel loading tips you actually have more control of your sample. The tip is designed to go into the gels with narrow wells. Gel tips allow dispensing your sample accurately and precisely without the risk of overflowing (and cross-contaminating) into the next well. Always run gel electrophoresis markers along with your samples since these are not only necessary to determine accurately protein sizes but also serve as a control for the separation.

Gel Temperature

Maintaining a constant temperature between 10^0C - 20^0C during separation is paramount in obtaining even migration of proteins. Otherwise, outer lanes will move slower than center lanes resulting in artifact known as smiling. To prevent this negative effect, you can ensure efficient heat transfer by completely filling the outer chamber of the electrophoresis apparatus with SDS running buffer and constantly stirring the buffer with a magnetic stirrer. Alternatively, heat can be reduced by lowering the running current during separation.

2D GEL-Two-dimensional gel electrophoresis with Coomassie

stained

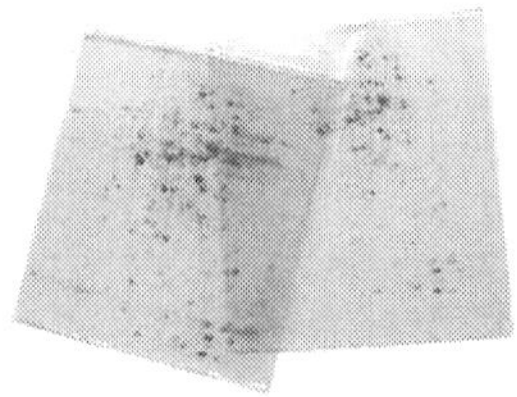

Robots are used for the isolation of protein spots from 2D gels in modern laboratories.

Two-dimensional gel electrophoresis, abbreviated as 2-DE or 2-D electrophoresis, is a form of gel electrophoresis commonly used to analyze proteins. Mixtures of proteins are separated by two properties in two dimensions on 2D gels.

Basis for separation

In the 2-D electrophoresis it separates the molecules perpendicularly from the first dimension and then creates an electropherogram in the second dimension. In the first dimension, molecules are separated linearly according to their isoelectric point. While in the second dimension, the molecules are separated at 90 degrees from the first electropherogram according to molecular mass. Since it is unlikely that two molecules will be similar in two distinct properties, molecules are more effectively separated in 2-D electrophoresis. The two dimensions that proteins are separated into using this technique can be based on the isoelectric point, protein complex mass in the native state and protein mass.

Separation of the proteins by isoelectric point is called isoelectric focusing (IEF). Here, a series of pH is applied to a gel and an electric potential is applied across the gel, making one end more positive than the other. If they are positively charged, the protein sample will be pulled towards the more negative end of the gel and if the protein sample is negatively charged it will be pulled to the more positive end of the gel. The proteins applied in the first dimension will move along the gel and will accumulate at their isoelectric point which is the point at which the overall charge of the protein is 0 (a neutral charge).

For the analysis on the functioning of proteins in a cell, the knowledge of their biochemistry is essential. Most often to be fully functional, all proteins will act together in a complexes system. The analysis of this proteins requires to be in the native state of the protein complexes. Therefore, native polyacrylamide gel electrophoresis (native PAGE) are used. In this PAGE, proteins remain in their native state and are separated in the electric field followed by their mass and the mass of their complexes respectively. To obtain a separation by size and not by net charge, as in IEF, an additional charge is transferred to the proteins by the use of Coomassie Brilliant Blue or lithium dodecyl sulfate stain. After completion of the first dimension then the complexes are destroyed by applying the denaturing SDS-PAGE in the second dimension, where the proteins of which the complexes are composed of are now separated by their mass.

Before separating the proteins by mass, they are treated with sodium dodecyl sulfate (SDS) along with other reagents. This denatures the proteins and binds a number of SDS molecules proportional to the protein's length. Because an unfold protein's length is roughly proportional to its mass. Since the SDS molecules are negatively charged, the result of this is that all of the proteins will have approximately the same mass-to-charge ratio with each other. In addition, proteins will not migrate when they have no charge, therefore the coating of the protein in SDS will allows the migration of the proteins in the second dimension. In the second dimension, an electric potential is again applied at a 90 degree angle. The proteins will be attracted to the more positive side of the gel proportionally to their mass-to-charge ratio. As previously explained, this ratio will be nearly similar to all proteins. The proteins separation will be slowed by frictional forces. The gel therefore acts like a molecular sieve when the current is applied, separating the proteins on the basis of their molecular weight with larger proteins being retained higher in the gel and smaller proteins being able to pass through the sieve and reach lower regions of the gel.

Detecting proteins

These proteins can then be detected by a variety of staining, but the most commonly used stains are silver and Coomassie Brilliant Blue staining.

Silver staining is 100x more sensitive than Coomassie Brilliant Blue with a 40-fold range of linearity therefore silver staining are more preferable to be used. In a silver staining, a silver colloid is applied to the gel to bind to the cysteine groups within the protein. The silver is darkened by exposure to ultra-violet light. The amount of silver can be related to the amount of protein at a given location on the gel. A dark gel is considered consist of a high amount of protein. This measurement can only give approximate amounts.

Common techniques

IPG-DALT

A common technique that can be utilized in the first dimension is through Immobilized pH gradient (IPG). This technique is referred to as **IPG-DALT**. In this technique, the sample is first separated onto IPG gel, then the gel is cut into slices for each sample which is then equilibrated in SDS-mercaptoethanol. This sample solution is then applied to an SDS-PAGE gel for resolution in the second dimension. However, it is not preferable for low molecular weight components as due to the loss of low molecular weight components during the transfer to the SDS-PAGE gel.

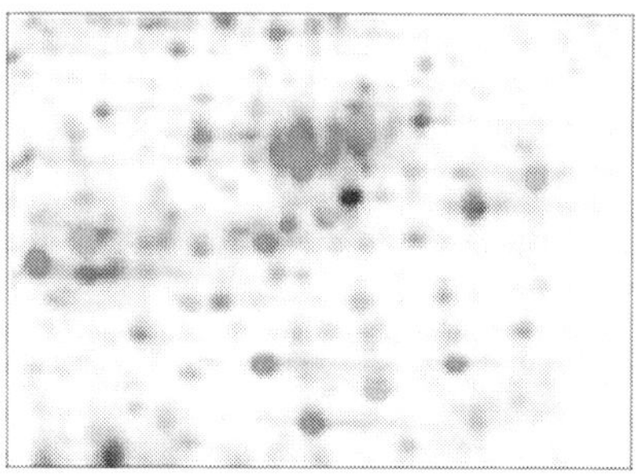

Warping: Images of two 2D electrophoresis gels, overlaid with Delta2D software. First image is colored in orange and second one is colored in blue. Due to running differences, corresponding spots do not overlap.

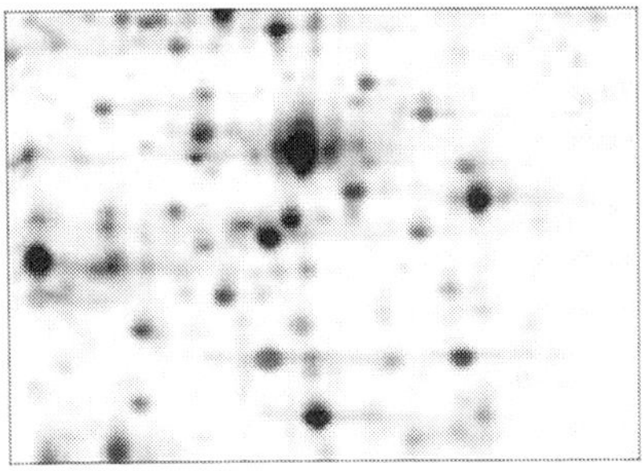

Common spots are colored black, orange spots are only present on the first image, blue spots are only present on the second image. In quantitative proteomics, these tools primarily analyze bio-markers by quantifying individual proteins and showing the separation between one or more protein "spots" on a scanned image of a 2-DE gel. Additionally, these tools match spots between gels of similar samples to show, for example, proteomic differences between early and advanced stages of an illness.

There are many software packages that is available for this analysis for example BioNumerics 2D, Delta2D, ImageMaster, Melanie, PDQuest, Progenesis, REDFIN, PDQuest and Progenesis. Challenges for automatic software-based analysis include those spots that are incompletely separated or overlapping, weak spots / noise, protein spot migrates to different positions on different gels, unmatched/undetected spots, mismatched spots, multiple spots may be erroneously detected as a single spot by the software will leads to wrong interpretation by the software.

Isoelectric Focusing

Isoelectric focusing (IEF) is a technique for separating different molecules by their isoelectric point (pI). Each molecules will have specific isoelectric point. IEF involves adding an ampholyte solution into immobilized pH gradient (IPG) gels. IPGs are the acrylamide gel matrix that co-polymerized with the gradient pH, which result in completely stable gradients except the most alkaline (>12) pH values. The immobilized pH gradient is obtained by the continuous change in the ratio of immobiline that is a weak acid or base defined by its pK value. A protein that is in a pH region below its isoelectric point (pI) will be positively charged and will migrate towards the cathode which is negatively charged electrode. As it migrates through a gradient of increasing pH, the protein's overall charge will decrease until the protein reaches the pH region that corresponds to its pI. At this point it has no net charge and so migration will be stopped. As a result, the proteins become focused into a stationary bands with each protein positioned at a point in the pH gradient corresponding to its pI. The technique is capable of extremely high resolution with proteins differing by a single charge being fractionated into separate bands.

Types of Isoelectric Focusing

Isoelectric focusing by living cells

Living eukaryotic cells perform isoelectric focusing of proteins in their interior to overcome the limitation of metabolic reaction by diffusion of enzymes and their reactants and to regulate the biochemical processes. By concentrating the enzymes of a particular metabolic pathways into a distinct and a small region of its interior, the cell can increase the rate of specific biochemical pathways by several orders of magnitude. By modification of the isoelectric point (pI) of molecules of an enzyme by either phosphorylation or de-phosphorylation, the cell can transfer molecules of the enzyme between different parts of its interior, to switch on or to switch off certain biochemical processes.

Microfluidic chip based isoelectric focusing

Microchip based electrophoresis is a promising alternative than capillary electrophoresis. Microfluidic has the potential to provide rapid protein analysis, straightforward integration with other microfluidic unit operations, whole channel detection, nitrocellulose films, smaller sample sizes and lower fabrication costs.

Multi-junction isoelectric focusing

The increased demand for faster and easy-to-use protein separation tools has accelerate the evolution of IEF towards in-solution separations. Here, a multi-junction IEF system was developed to perform fast and gel-free IEF separations. The multi-junction IEF system utilizes a series of vessels with a capillary passing through each vessel. Part of the capillary in each vessel is replaced by a semipermeable membrane. The vessels contain buffer solutions with different pH values. This buffer functioned to establish an effective pH gradient inside the capillary and together with an electrical contact that connected to a high-voltage power supply divider. It will establish electrical field along the capillary. When a sample of a mixture of peptides or proteins is injected into the capillary, the presence of the electrical field and the pH gradient separates these molecules according to their isoelectric points similar mechanism if gel were to be used.

Microarray

Microarray is a kind of gene chip technology. It is a 2D array on a solid substrate that assays large amounts of biological material using high-throughput screening with parallel processing and detection methods. The "gene chip" industry started with the establishment of companies, such as Affymetrix, Agilent, Applied Microarrays, Illumina and others. The technology of DNA microarrays has become the most sophisticated and the most widely used than the use of protein, peptide and carbohydrate microarrays.

Types of microarrays include:

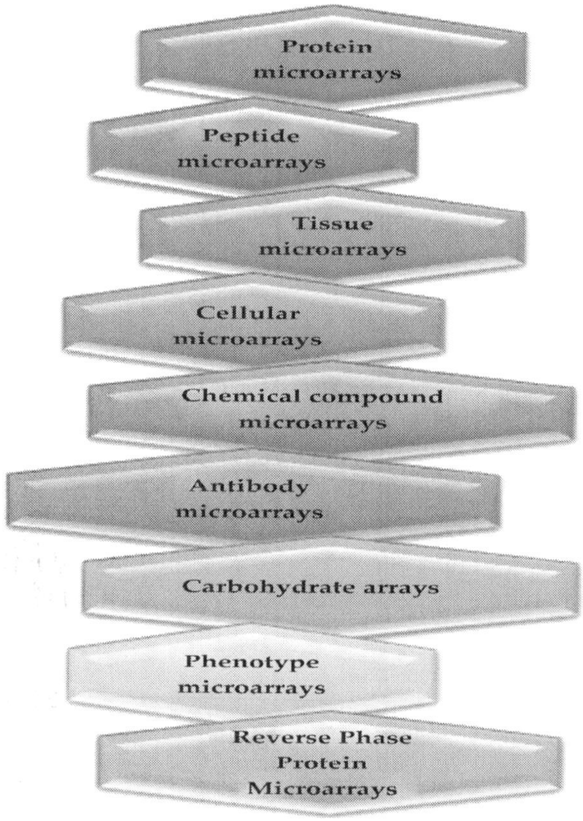

Immunoassay; Bead-Based Immunoassays

Western blot is a technique that is very useful for protein detection as it allows the user to quantify the protein expression as well. and Western blot can be seen as an intricate balance, as the researcher attempts to get a nonspecific, yet strong signal.

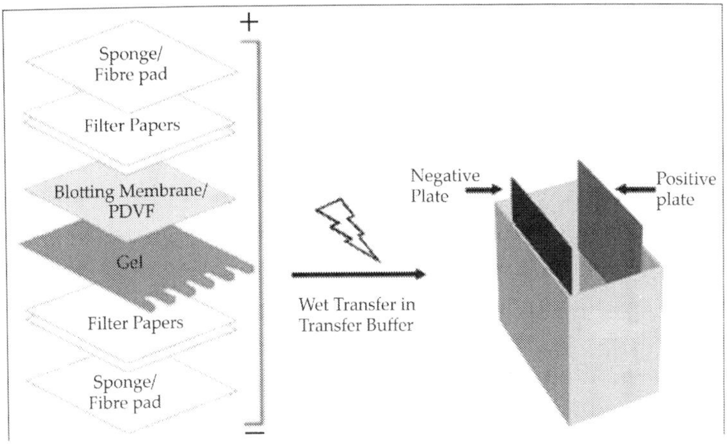

TIPS AND TECHNIQUES- Immunoassay

Protein sample will be used for immunoassay analysis. Protein sample can be extracted from different kind of part such as cells, tissue, plant stem, leaves or flower or from animal`s organs. All samples are required to be wash in cold phosphate buffered saline (PBS) and left rocking gently. Fresh PBS are added to dislodge the cells and the mixture is transferred into microcentrifuge tubes. The sample are allowed to be centrifuged at 1500 rpm for 5 minutes and the supernatant are discarded after the completion of centrifugation. 180 μL of ice cold cell lysis buffer with 20 μL of fresh protease inhibitor cocktail are prepared. In case a protein concentration is not high enough at the end, it is advised to repeat the procedure with a higher proportion of protease inhibitor cocktail. Followed with that, a 30 minutes on ice incubation was carried out and then it need to clarify the lysate by spinning for another 10 minutes at 12,000 rpm at 4°C. The supernatant (or protein mix) will be transferred into a fresh tube and store on ice or frozen at -20°C or -80°C. The concentration of protein will be measured using a spectrophotometer. By using the concentration, sample preparation will be calculated using concentration= mass/volume. 5 μL of sample buffer will be added to the sample to make the volume in each lane are equalized using double distilled H_2O (dd

H_2O). This sample solution will be mixed well and heated with dry plate for 5 minutes at 100°C. After preparing the 10% stacking gel solution, the rack for gel solidification are assembled. 10% AP and TEMED will help in solidifying the solution; there fore, both gels can be prepared at the same time and the AP and TEMED reagents are only added at the end. Add stacking gel solution carefully using suction pipette until the level is equal to the green bar holding the glass plates follow by adding water to the top. Wait for 15–30 minutes until the gel turning solidified prior adding the separation solution. The separation gel was added after removing the water layer by tilting the whole apparatus and with a paper towel the water can be easily removed. After adding the separation gel, insert the comb and ensure that there are no air bubbles. Wait for the gel to be solidified completely and it can be easier checked by leaving some gel solution in a tube. For electrophoresis, pour the running buffer into the electrophorator. Place the gel inside the electrophorator and connect to the power source red to red and black to black. Make sure the buffer covers the gel completely and remove the comb carefully. In each lane load marker and sample according to what was decided and recorded in the log book. The first lane always will be marker (6 μL) followed by samples (15 μL) in to each well. Run the gel with low voltage (60 V) for separating gel and use higher voltage (140 V) for stacking gel. Run the gel for approximately an hour, or until the dye front runs off the bottom of the gel. In the meantime user can prepare the filter sheets. Cut 6 filter sheets and one polyvinylidene fluoride PDVF membrane with the same dimensions of the prepared gel. The membrane need to be wetted, PDVF membrane will be wetted in methanol, the sponge and filter paper will be wetted in transfer buffer. Once the separation was completed separate the glass plates and retrieve the gel gently to avoid breakage in the gel. Users need to create a transfer sandwich as follows:

- Sponge
- 3 Filter Papers
- Gel PVDF
- 3 Filter Papers

User need to ensure that there is no air bubbles or extra liquids between the gel and PVDF membrane. Placed the sandwich to the transfer apparatus, which should be placed on ice to maintain 4°C. Now, add transfer buffer to the apparatus, until the sandwich is covered with the buffer completely. For transferring, place electrodes on top of the sandwich, ensuring that the PVDF membrane is between the gel and a positive electrode for 90 minutes. In case, the thickness of the gel is 0.75mm the running time will be around 45 min as the thickness of the gel is proportional to the running time. During blocking and antibody incubation, user will block the membrane with 5% skim milk in TBST for 1 hour. After blocking, add primary antibody in 5% bovine serum albumin (BSA) and incubate overnight in 4°C on a shaker. Once incubation was completed, wash the membrane with TBST for 5 minutes for 3 times. All washing and antibody incubation steps should be done on a shaker at room temperature to ensure even agitation. Add secondary antibody in 5% skim milk in TBST and incubate for 1 hour. Wash the membrane with TBST for 5 minutes for 3 times. Prepare ECL mix by following the proportion of solution A and B provided by the manufacturer and ensure that the solution covers the top and bottom of the membrane during the incubation which need to be carry out for 1–2 minutes. Finally, the user can visualize the result in the dark room.

TOOLS AND TECHNIQUES USED IN PLANT TISSUE CULTURE

In this sections we will discussing and introducing types of tools and techniques that usually use for plant tissue culture research work.

pH Meter

pH meter is used to measured pH of the prepared media and solution. Usually tissue culture media is prepared at pH5.0-6.0 and pH solution may be vary.

Measurement of pH:

In this day, pH of a solution was measured by the pH probe. It measures pH of the hydrogen ions surrounding a thin walled glass bulb at its tip. The probe produces a small voltage that is measured and displayed as pH units by the meter. The meter circuit is fundamentally no more than a voltmeter that displays measurements in pH units instead of volts.

For example:

(i) At neutral pH (pH 7) the voltage at the probe's output is 0 volts.

(ii) At alkaline pH, the voltage at the probe's output ranges from + 0 to + 0.41

(iii) At acid pH, the voltage at the probe's output ranges from – 0.41 volts to – 0.

Applications:

- pH meter is used to adjust pH of different solutions, during preparation of buffers and culture media or pH of the medium in the bioreactor.
- It is also widely used to determine the pH of cells (cell sap).

Methods Of Sterilization

Asepsis

Asepsis is the state of being free from bacteria, viruses and other microorganisms. Plant tissue culture requires contamination free environment, tools and samples or strict maintenance of germ free system in all the operations. The basic rules and practices of asepsis are followed in all the tissue culture laboratories. Plant tissue culture media are rich in nutrients and very suitable for the growth of microbes. These microorganisms will grow faster by consuming the nutrients and at the same time suppress the growth of plant tissues. This unwanted growth of microbes is referred as contamination. There are many ways by which these microbes suppress the growth of plant cells and tissues such as by releasing of enzymes and toxins (which inhibit the growth of plant cells), selective absorption of specific nutrients and some tissue infection. To achieve success in cultivation of higher plant parts, it is essential to avoid these contamination to save cultures. This has led to the development of 'clean area concept' where all the area, tools and working places should be free from microbes.

- Minimize the external air flow around the working area so it is possible to avoid spores of microorganisms to move in along with the air currents over the sterile areas.
- Basically, a fan should not be used in laminar air flow bench room (inoculation room).
- All tissue culture lab should be air-conditioned.
- All the sterilized media, nutrients and tools need to be stored properly in clean cabinet.
- Designate separate area for cleaning and washing and for the preparation of medium.
- UV tubes lights should be on before inoculations.

Surface Sterilization of Explants:

In tissue culture, most contamination are tend to be introduced through the explant because of lack of complete/proper sterilization. It can be fungal or bacterial. This type of contamination which can be seen in a few days prior to culturing can be a very difficult problem.

The contamination will produce "ooze" on solid medium and turbidity in liquid cultures. If the contamination is from fungi, it will look "furry" on solid medium and often accumulate in little balls in liquid medium. In case the contamination is derived from the plant samples they may survive surface sterilization because they are in interior tissues.

There are various solution that can be used for sterilizations

1. 1% solution of sodium hypochlorite (commercial bleach).
2. 7 % saturated solution of calcium hypochlorite.
3. 1 % solution of bromine water.
4. 70% ethyl alcohol.
5. 0.2% mercuric chloride.
6. 10% hydrogen peroxide solution.
7. 1% silver nitrate solution.

Function Of Disinfection Solution

The various metallic ions can be arranged in a series of decreasing antibacterial activity. Hg^{2+} and Ag^{+}, are effective at less than one part per million (ppm), because of their high affinity for sulfhydryl group. Bacteria are killed by Ag containing 10^5 to 10^7 Ag^{+} ions per cell. The concentration required for killing is markedly affected by inoculum size.

Chlorine was used to prevent transmission of puerperal sepsis by the physician's hand. Chlorine combines with water to form hypochlorous acid (HOC1) which act as a strong oxidizing agent.

Sodium hypochlorite (NaOCl) solutions are (200 ppm chlorine) are used to sanitize clean surfaces. Sodium hypochlorite is commercial bleach available in the market as 3-6% solution and can be used directly. Sometimes users can dilute the solution of sodium hydroxide. Powdered bleach is also added into water supply for killing microbes.

Process of sterilization during tissue culturing

1. First user need to clean the working area with ethanol and start the air flow of the laminar air flow bench.
2. The sterilized petri dishes, distilled water, scalpel, filter paper sheets alcohol, disinfectant solution, forceps, bead sterilizer or beaker or coupling jar, spirit lamp or gas burner are put on the UV light of the bench for 30 minutes.

3. Clean the working area and hands with alcohol, put on mask and cap and light the spirit lamp.
4. Now the plant material is brought into the laminar air flow for further sterilization prior to culturing process.
5. Keep 3-4 petri dishes in a line containing disinfectant in 1st plate and autoclaved distilled water in subsequent plates.
6. Place plant pieces in 1st plate and immerse the material with the help of forceps for 5-10 minutes depending upon the disinfectant used.
7. Transfer material from 1st to 2nd plate, rinse gently and pass to 3rd and 4th plate, one by one with thorough rinsing.
8. Finally, drain the distilled water or place the material in a fresh petri dish or filter paper and prepare suitable sized explants.

Essentially, method for sterilization is different for living materials than tools and glassware. Now a day's many items are available as ready to use, pre-sterilized (by gamma radiation) and disposable. Uses of such disposable materials have ease the tissue culture research.

There are three types of sterilization that can be applied:

1. Dry heat
2. Wet heat
3. Filter sterilization.

1. Dry heat:

Dry heat sterilization can be applied to glassware and metal tools which do not get charred by high temperature. All this items are put in containers or wrapped in aluminum foil prior to placing it in dry oven for sterilization for a period of not less than three hours at a temperature of 140-160 °C. However this techniques are not suitable for media and plastic ware.

2. Wet heat:

Wet heat is the most popular method of sterilization for both tools and media. This method involves of autoclaving at 121°C with a pressure of 15 psi (pounds per square inch) for 15 min (1.02 kg/cm^2). Modern autoclaves are capable of providing saturated steam treatment ranging from 70-132 °C, which is a pressure of up to 25 psi. All the vessels containing medium should be placed vertically and should not be filled more than 40% to their total capacity, to be save, the cap is not overly tighten.

3. Filter sterilization:

Filter sterilization or cold sterilization is used when a solution or medium cannot be sterilized by autoclaving. It is the property of the filter (porosity 0.22 to 0.45 µm) to retain the entire microorganism and make the solution free from microbes. This exclusion of microorganisms makes the solution sterilized without heating or autoclaving.

Usually only heat sensitive liquids such as plant growth regulators (IAA, GA_3, Zeatin, Abscisic Acid), some bio-chemicals, enzymes for protoplasts isolation, antibiotics or nutrient solution can be sterilized by this method. This type of sterilization will avoid decomposition during autoclaving. After sterilization, solution is added to the medium.

The filters such as glass and membrane filters (nitro-cellulose membranes) are commercially available. Glass (sintered glass filters, Borosil) are available in different grades (G-1 to 5) and the G-5 (1-2 pm) is used for bacteriological purposes. These filters have a porous glass disc for filtration of liquids and gases as a filter media which is non-corrosive and reusable.

Membrane type filters of different sizes and porosity (0.22 to 0.45 µm) are available as disposable or reusable filters. These filters can be connected through tubing or syringe to pass the solution. Non-sterilized filters are wrapped in aluminum foil or paper to be autoclaved at 121°C for 15 min and then taken to laminar air flow bench. Solution is placed in funnel or syringe and allowed to pass through filter under suction (glass) or pressure (membrane).

A membrane filter or glass tube, filled with cotton may be placed between vacuum flask and suction pump to avoid entry of air from the outside during operation. Sterilized filtrate is collected in the flask and dispensed in the liquid or medium with the help of sterilized pipette fitted with cotton plug. Volume added to the medium is adjusted to reach a final concentration. Membrane type filters are also used as inlets and outlets of bioreactor system and to draw samples.

There are a variety of wet and dry heat treatments, radiations, filtration, gas and chemical treatments which available for direct sterilization of material. Gas treatments are rarely used in the laboratory. Ultraviolet light treatment of working surfaces and sterile rooms are used in the labs while gamma radiation is used for the preparation of pre-sterilized disposable plastic-wares.

Other Type Of Sterilization:

Surgical blades and scalpels are not sterilized by dry heat because the high temperature makes the cutting edge dull. Therefore, they are usually immersed in 80% v/v (volume by volume) ethyl alcohol until required and sterilized during usage by frequent immersion in alcohol and flaming.

Usage of formaldehyde, sulphur some time is needed to fumigate lab area from microorganisms. The fumigated place should not be used for 3-4 days. Culture rooms or growth rooms are fumigated only after removal of cultures.

Medium and Its Preparation:

Murashige and Skoog's medium (Murashige and Skoog, 1962) is a landmark in plant tissue culture research and is the most frequently used medium for all types of tissue culture work. Based on its composition, other media were evolved to meet the diverse experimental and species-spccific rcquirement.

The methodology of plant tissue culture has advanced to the stage, where tissues from any plant species can be cultured successfully with efficient media formula. All the media consist of mineral salts, a carbon source (generally sucrose), vitamins and growth regulators. The cells of most plant species can be grown on completely defined media.

Medium Composition:

The nutrient medium for most plant tissue cultures is comprised of five groups of ingredients

a) Inorganic nutrients

b) Carbon source

c) Vitamins

d) Growth regulators

e) Organic supplements

Inorganic Nutrients: consist of macro- and micro-elements as salts. Usually nutrient media contain 25 mM each of nitrate and potassium. For regular culture and cell cultures, they combined nitrogen and ammonium. Ammonium is essential in lower concentration for most cultures. A concentration of 1-3 mM is always adequate for calcium, magnesium and sulphate. The required micronutrients include I, B, Mn, Zn, Mo, Cu, Co and Fe.

Carbon Source: Glucose, fructose, maltose or sucrose are type of carbon source but sucrose is the preferred source for most of the cultures. Sucrose will rapidly converted into glucose and fructose.

Vitamins and amino acids:

Thiamine, pyridoxine and nicotinic acid are commonly used as vitamins in B_5 and as MS media. Amino acids and organic supplements serve as source of reduced nitrogen. In case of inadequate nitrogen, complex organic nitrogen supplement like casein hydrolysate (0.1-1 g/L) can be supplemented. Glycine is commonly used amino acid. Other organic supplements are such as coconut milk, yeast extract, peptone and malt extract. However, synthetic media are preferred and organic supplement of unknown chemical nature is used only when it is essential.

Plant Growth Regulators (PGR):

A balanced combination of PGR is required for sustained growth of a plant. Usually two types of combinations are used;

- for cell proliferation consisting of :2,4- dichlorophenoxy acetic acid (2,4-D); 1-naphthalene acetic acid (NAA); cytokinin (kinetin, benzyl adenosine; 2-isopentyladencisine; zeatin, thidiazuron),

For regeneration essentially containing low auxin [(NAA, IAA, Indole butyric acid (IBA)]; a cytokinin in high amount, but not 2,4-D known to induce cell proliferation but suppresses differentiation in dicot plants. However, 2,4-D and 2,4,5-T (2,4,5-trichlorophenoxy acetic acid) are effective in inducing somatic embryogenesi.

Medium Preparation:

A convenient approach to prepare a tissue culture medium is to have stock solutions of all the nutrients in a 10x or 50x concentration. Desired medium is prepared by suitably diluting the appropriate amount of stock solutions. It may be advantageous to have separate stock solutions of calcium salt and potassium iodide.

Once all the ingredients are mixed, sugar will be added and pH is adjusted to 5.8-6.0. All the ingredients should be thoroughly mixed before dispensing in the vessels. All the vessels are plugged with non-absorbent cotton, covered with aluminum foil and autoclaved at 121 °C for 15 min. Prepared media can be stored for a few weeks before inoculation.

Autoclave

An autoclave have a body, an internal (or external for small sized autoclaves comparable to household pressure cooker) heating system, a container to hold material, its cover fixed with pressure gauge, safety valve, pressure release valve etc. Lid is tightened with the help of screws and a gasket seals the body and lid. A jacket, paddle lifter, timer and indicator etc., are also provided with large sized autoclaves. Autoclaves may be constructed of aluminum , mild steel, stainless steel or gun metal. Industrial autoclave can accommodate large trolley containing huge number of glassware's or large bioreactors. Autoclave machine is used to sterilize medium, glassware and tools for the purpose of plant tissue culture. Sterilization of material is carried out by increasing moist heat at 121 °C due to increased pressure inside the vessel (15-22 psi, pounds per square inch or 1.02 to 1.5 kg/cm^2) for 15 minutes for routine sterilization. Moist heat kills the microorganism and makes the material free from microbes.

Techniques

Place the materials either

- Wrapped with aluminum foil or paper or in metal box or in glass container
- Glass-wares containing medium were closed loosely in the bucket.

Check water level for appropriate level, tighten all the screws and switch-on. Allow the steam to pass freely from release valve for 5 minutes and then close the valve.

After attaining a pressure of 15 psi, timed 15 minutes for sterilization and then switch-off. Pressure is maintained by safety valve. Modern autoclaves are fitted with temperature and time control units and can automatically control the period for sterilization and then switch-off themselves.

Empty vessels, beakers, graduated cylinders, etc., should be closed with a cap or aluminum foil. Tools should also be wrapped in foil or paper or put in a covered sterilization tray. It is important that the steam penetrate the items in order for a complete sterilization.

For Large Sized Vessels And Large Volume Flasks Containing High Amount Of Liquid, Duration For Sterilization Should Be Increased Accordingly. The Vessels Should Not Be Filled More Than 1/3 Of Its Capacity.

For Proper Sterilizationtips

1. Ensure that water level is submerged the heating elements.
2. Check the spring of safety valve and keep the opening valve clean.
3. Tight the both screws of the lid simultaneously. But ensure not over tighten the screws to avoid damage to the gasket.
5. Use permanent marker to mark the flasks and its medium.
6. All the electrical equipment should be properly earthen to avoid electric shock.

Plant Growth Chamber:

Plant growth chambers can be constructed in a suitable sized room or can be purchased as commercially available equipment.

Plant growth chamber consist of various types of environmental control systems with sensors and timers to monitor the desired values:

- Light-intensity
- Duration cycle control.
- Temperature control and regulation.
- Humidity control and regulation.

Light Of The Future: Led

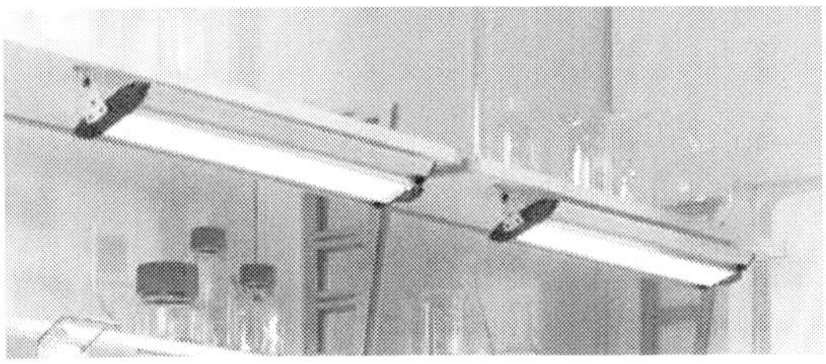

Laboratory work is very precise and detailed in nature and lighting can be one of the most important tools of all. Effective lighting is crucial for employee health and safety. Light output is generated along the entire length of the luminaire without any dark areas. High-quality diffusers control glare and reflections on glossy surfaces. Subtle, evenly-dispersed, flicker-free light reduces eye strain and improves productivity. The high color rendering index (cri) of 85 (with the clear, prismatic lens) and 90 (with the translucent lens) generates superb color reproduction. Whatever the version, LED is always energy-efficient, consuming 30% less energy when compared to conventional task luminaires with the same luminous power. The smaller, 1400 lux model uses only 15w of energy consumption, while the larger 2600 lux and 3600 lux versions use only 31w and 47w respectively, including the power supply. Since the development of high-brightness blue and white LEDs, the market for led light has been booming.

Temperature

In modern equipment, temperature is precisely regulated by temperature sensor made from platinum. Temperature setting is very important in growth chamber. In all cases, air conditioning units provide the cooling. However, it is always advisable to keep a spare compressor unit to avoid delay in repairs that can leads to damage in cultures. Usually temperature of 22-28 °C is used for growing plant tissue culture chamber. Temperatures should be measured in a constructed growth chamber at different levels and places in order to have a correct temperature setting.

Humidity

Humidity inside the growth chamber is provided by humidifier and controlled by humidistat. Usually 60% relative humidity is used to maintain healthy growth. Low relative humidity may cause early drying of medium while high humidity may cause fungal growth in the environment and on a various particles. Thus, in a growth chamber, light, temperature and humidity are precisely controlled so that cultures are grown in a controlled environment. All the controls are set on control panel.

Laminar Air Flow Bench:

Laminar air flow bench is the main working table for sterilized manipulations related to plant tissue culture. This equipment fitted with High Efficiency Particulate Air (HEPA) Filters, which allow air to pass but retain all the particles and micro-organisms. These HEPA Filters have a very small pore size (0.3 μm) with 99.97-99.99% efficiency.

Air blown by a blower is passes through these filters and a positive pressure of air is maintained from inside to outside. This positive pressure does not allow any particle to enter in the working area of the bench. Air pressure inside the instrument is measured by a manometer and pressure of more than 12 bars shows choking of filters and at this point filters should be replaced. This equipment is fitted with UV light and visible light source.

UV is switched-on for 30 minutes before starting the work to make area free from microbes. Laminar air flow bench made of steel and wooden cabinets is available with different working table size and for vertical (downward air flow) or horizontal (horizontal flow) model. HEPA filters are also used for other purposes to create 'clean area' for culture rooms and inoculation room. If laminar air flow bench is placed in such a clear area, efficiency and life of the equipment are increased.

Thermometer

The thermometer is a device that measures temperature or temperature gradient using a variety of different principles. A thermometer has two important elements: the temperature sensor (e.g., the bulb on a mercury thermometer) in which some physical change occurs with temperature, plus some means of converting this physical change into a value (e.g., the scale on a mercury thermometer). For the working temperature range, the meniscus or interface between the liquid is within the capillary. With increasing temperature, the volume of liquid expands and the meniscus moves up the capillary. The position of the meniscus shows the temperature against an inscribed scale. Since these are transparent, the liquid is made more visible by the addition of a red or blue dye. One half of the glass containing the capillary is usually enameled white or yellow to give a background for reading the scale. During the dark period, temperature remains slightly (1-2 °C) lower than light period (all bulbs and tube lights of the culture room increases the temperature).

Hygrometer

Hygrometers are instruments used for measuring humidity. A simple form of a hygrometer is specifically known as a “psychrometer” which consists of two thermometers, one of which includes a dry bulb and the other of which includes a bulb that is kept wet to measure wet-bulb temperature. Evaporation from the wet bulb lowers the temperature, so that the wet-bulb thermometer shows a lower temperature than that of the dry-bulb thermometer, which measures dry-bulb temperature.

Relative humidity is computed from the ambient temperature as shown by the dry- bulb thermometer and the difference in temperatures as shown by the wet-bulb and dry-bulb thermometers. Relative humidity can also be determined by locating the intersection of the wet- and dry-bulb temperatures on a psychrometric chart.

Dial type hair hygrometer is a convenient tool to measure humidity in the culture room. Humidifier is used to maintain 60% relative humidity (RH) in the culture room. Humidistat is a bimetallic thermocouple device to control and regulate the function of humidifier to maintain the humidity. Distilled water should be filled in the humidifier.

The RH present in the culture room is measured by hair hygrometer. As the name suggests, a chemically treated hair elongates with increased humidity and shortens with dryness. It is calibrated from 0 to 100% RH. At lower humidity, medium dries rapidly whereas at higher humidity chances of fungal growth over all surfaces and cotton plugs is increased. Therefore, about 60% RH is maintained in the culture room.

LUX Meter

Light intensity can be measured by photometer or lux meter. A photometer consists of photoelectric cell and a micro-ammeter. Photoelectric cell is sensitive for light and converts light into current. Micro-ammeter shows reading due to this current and its needle moves. Nowadays digital read out is given by appropriately converting the current into digital signal. High intensity is proportional to the current generated in the photoelectric cell by falling light. The unit of illumination is called as ‘lux’ and the scale of ammeter is calibrated in lux. Usually 2000 to 2500 lux is provided to the cultures maintained in light in the culture room. This instrument is a sensitive tool and need to be handled with care. It should not be exposed to the sunlight without switching the proper reading switch to high light illumination.

Light is the form of radiant energy, i.e., electromagnetic radiation of specific wavelength. Visible light as we perceive, is located in narrow wavelength region of spectrum between 380 to 760 nm. Light has dual characters, displaying both wave properties (refraction, diffraction, interference and polarization phenomena) and particle properties (light is radiated in discrete amounts of energy or photons).

Properties of light should be defined either as irradiance or an illuminance. It is to note that an irradiance measurement is not spectrally defined, whereas, an illuminance measurement indicates the level of visible light as the human eye would see it.

Artificial light is provided by cool, white, fluorescent tubes and/ or incandescent bulbs of different classes of artificial light sources, fluorescent sources have been used almost exclusively for plant tissue culture because they are more efficient at producing broad band visible light than incandescent bulbs and are available in low output wattage than lamps.

TOOLS AND TECHNIQUES IN ANALYZING BIOCHEMICAL COMPOUNDS

Colorimeter

The most commonly used method for determining the concentration of biochemical compounds is colorimetry. In colorimetry a lamp is usually used as the source of radiation. The wavelength of light depends upon the quality of the lamp source. It uses the property of light such that when white light passes through a colored solution, some of the wavelengths are absorbed more than others. The intensity of the color is directly proportional to the concentration of the compound being measured. In another words, the amount of light absorbed is proportional to the intensity of the color and therefore, to the concentration. The big advantage is that complete isolation of the compound is not necessary.

The relationship between concentration of solution and light it absorbs is given by the following laws.

Lambert's Law

When a ray of monochromatic light passes through an absorbing medium, its intensity decreases exponentially as the length of the absorbing medium increases.

Beer's Law

When a ray of monochromatic light passes through an absorbing medium, its intensity decreases exponentially as the concentration of the absorbing medium increases.

Transmittance

The ratio of intensities is known as the transmittance and this is usually expressed as percentage. Absorbance or extinction is known as the optical density of the substance. Absorbance increases with increase in concentration of the solution.

Detection of Radiation

Detection wavelength is selected on the basis of wavelength of light coming out of a solution while absorption of a particular wavelength of light is basically refer to the characteristic of the compound.

Applications

Measurement of concentrations:

Spectrophotometers are widely used for quantitative measurements of substance concentration in the solutions by measuring absorbance at the optimal wavelength.

Absorption spectra:

Identification and structure evaluation of various isolated pure compounds can be done by UV-Visible spectrophotometry. All compounds have very specific absorption spectra depending on the molecular structure. Comparison of spectra helps in identification of type of the compounds.

Enzyme kinetics:

In enzymatic reactions, changes in absorbance are recorded to determine the reaction velocity and concentration of the product formed/substrate utilized. Current spectrophotometers are coupled with a microprocessor or computer which allow the storage of data and figures more systematically.

Centrifugation:

A centrifuge is an instrument which produces centrifugal force by rotating the samples around a central axis with the help of an electric motor. Centrifuges can be categorized as the

- clinical type (5-10,000 rpm)
- refrigerated high-speed centrifuges (10,000-20,000 rpm)
- ultra -centrifuges (20,000 to 80,000 rpm)refrigerated centrifuge

All the enzymatic work requires refrigerated centrifuges to keep the samples cool during centrifugation and to protect enzymes from inactivation by heat.

Centrifugation with heating and vacuum (suction) is used for concentrating extraction (instrument called sample concentrator) and rapidly removing solvents. There are two types of rotors, angle head and swing.

There are 2 types of Ultracentrifuges:

Analytical Model:

This consists of rotors and tubes, called cells. The instrument is designed to allow the operator to follow the progress of the substances in the cells, while the process of centrifugation is in progress. By estimating sedimentation velocity during the process, the molecular weight, purity etc. can be determined.

Preparative Model:

This is used for purification of the components of macromolecules or other substances and all determinations are made at the end of centrifugation. The instrument has no monitoring device, while large centrifugal forces are set for a fixed time period. Centrifugation is the most widely used technique for separation of various metabolites and also used to separate non-miscible liquids during extraction of secondary metabolites, e.g., water (aqueous), chloroform (organic) mixture.

A wide variety of centrifuges are available, ranging in capacity and speed. During the process of centrifugation, solid particles experience a centrifugal force, which pulls them outwards, i.e., away from the center. The velocity with which a given solid particle moves through a liquid medium is related to angular velocity. The principle of centrifugation is that, an object moving in a circular motion at an angular velocity is subjected to an outward force (F) through a radius of rotation (r) in cms, presented as $\omega^2[F = \omega^2 r]$.

F frequently expressed in terms of gravitational force of the Earth, commonly referred to as RCF (Relative centrifugal force) or g by the following formula [RCF = $\omega 2r / 980$]. The ω^2 operating speed of the centrifuge is expressed as revolutions per minute 'rpm', which can be converted to radians.

Applications:

Centrifugation is widely used in

- analytical techniques,
- preparation of extracts,
- separation of non-miscible liquid mixtures,
- purification of enzymes, inhibitors and removing particles (silica, etc.) from the solvents.

Chromatography:

Chromatography is a technique to separate molecules on the basis of differences in size, shape, mass, charge and adsorption properties. The magnitude of these interactions depends upon the particular method used. Usually column chromatography is used to separate large quantities of compounds, whereas, paper chromatography (PC) or thin-layer chromatography (TLC) in one or two dimensions, is used for analytical work.

The mobile phase can be a gas or a liquid, whereas the stationary phase can only be a liquid or solid. In liquid column chromatography (LCC), separation involves a simple partitioning between two immiscible liquid phases, one stationary and the other mobile, the process is called liquid-liquid (or partition) chromatography (LLC). When physical surface forces are mainly involved in the retentive ability of the stationary phase, the process is denoted liquid-solid (or adsorption) chromatography (LSC).

In ion exchange chromatography (IEC), ionic components of the sample are separated by selective exchange with counter ions of the stationary phase. Use of exclusion packing's as the stationary phase brings about a classification of molecules based largely on molecular geometry and size.

Paper Chromatography:

In paper chromatography paper sheets derived from cellulose is used. It makes an ideal support medium where water is absorbed between the cellulose fibers and forms a stationary hydrophilic phase. The concentrated mixture is spotted onto the paper, dried with a hair-drier and the chromatogram is developed by allowing the solvent to flow along the sheet. The solvent front is marked and after drying the paper, the positions of the compounds present in the mixture are visualized by a suitable staining reaction. The R value measured by the ratio of the distance moved by the compound to that moved by the solvent system. For solvent system that was controlled in terms of concentration, temperature and pH normally will have a constant pH.

R_f value is defined as [R_f = the distance moved by solute / the distance moved by solvent front].

Alcoholic extracts of plant material with or without partial purification are used for chromatography.

Biological materials should be desalted before chromatography by electrolysis or electro-dialysis.

Excess salt results in a poor chromatogram with spreading of spots, it can also affect the chemical reactions used to detect the compounds being separated and leads changes in their R values. The sample (10-20 μl) is then applied to the paper with a micropipette or capillary.

Separation using Paper

The most frequently used paper for separation is known as Whatman No.2, Whatman No.3, Whatman Nos. 4 and 5. These papers differ in their thickness. The papers are impregnated with a buffer solution before use or chemically modified by acetylation. Ion exchange papers are also available commercially. For the separation of lipids and similar hydrophobic molecules, silica-impregnated papers are available commercially.

Solvent

Choice of solvent always depends on the experimental sample. If the compound is move close to the choice of solvent it is consider that the compound is too soluble. While if the compound is crowded around the origin it is consider the compound are not soluble to the solvent. Therefore, the choice of the solvent will be the solvent that can be solubilize all the compounds and not solubilizing the compound. This will lead the R_f values of the components of the mixture are spread across the length of the paper. Once the desired solution was selected it is advice that the developing chamber should be saturated with solvent, by using filter paper lining inside the chamber.

2D- Chromatography

In 2D chromatography the compound is separated in the first solvent, which is volatile, then in the second solvent. In between both solvent the chromatography paper is dried and turned into 90 degree when merged in the second solution. After separation, a map is obtained and compounds can be identified by comparing their position with a map of known compounds developed under the same conditions.

Detection of spots

The spot will be detected by specific reagents. The reagent is sprayed under fume hood or rapidly dipped it in a volatile solution. Sometimes viewing under ultraviolet light is also useful since some compounds, which absorb strongly show up as dark spots against the fluorescent background of the paper.

Thin Layer Chromatography:

Principle

Thin layer chromatography principle is similar to other paper chromatography but user can apply a variety of supporting media by adsorption ion exchange, partition chromatography, or gel filtration. Due to its thin concept it can be completed within an hour, compounds can be detected at a lower concentration and separated compounds can be detected by corrosive sprays. There are now a number of prepared thin layer plates using different adsorbents on various supporting materials such as glass, plastic and aluminum that are available commercially and these may be more convenient to use than trying to prepare plates in the laboratory.

Calcium sulphate is sometimes incorporated into the adsorbent to bind the layer to the plate and, because of this it is advisable to work rapidly once the adsorbent is mixed with water. It is essential to make sure that the atmosphere of the separation chamber is fully saturated with the solvent mixture, otherwise R_f values will vary widely from tank to tank. However, horizontal chambers of high performance TLC (HPTLC) are very useful and convenient as they require less time, solvent and no stabilization time. Development of the plate is usually by the ascending technique and is very rapid.

TOOLS AND TECHNIQUES IN GENE ANALYSIS

Molecular Biology

There are various tools and techniques in gene analysis that can be applied based on applications. In the next sections, specific tools based on application were explained in detail. In molecular biology, DNA sequencing is used to study genomes and the encode protein. Information derived through this tool allows researchers to identify changes in genes, associations with diseases and phenotypes and identify potential drug targets. The DNA sequencing used in molecular biology can be manipulated and translate it into a RNA strand. DNA strands are considered a template or a coding strand. The template is the only sequence that can be translated into the RNA sequence. The DNA strand is transcribed into a RNA strand, which is then used to create a poly-peptide.

Evolutionary Biology

DNA sequencing is applied in evolutionary biology because DNA is an informative macromolecule transmitted from one generation to another. The result of DNA sequencing will uncover the relationship between organism and how they evolved.

Metagenomics

In metagenomics studies, the sequencing enables researchers to determine the type of microbes present in a microbiome, identified the organisms present in a body of water, sewage, debris filtered from the air, or swab samples from organisms. This result is important for research in all fields.

Key strength

- Enables bacteria identification and comparison within a mixed population sample
- Accelerates discovery of new SNP markers that translates to virulence or antibiotic resistance
- Promotes accurate evaluation of microbial biodiversity within a given sample
- Contributes to microbial phylogeny and taxonomy

Phylogenetic And Taxonomic Microbiology Studies

In phylogenetic and taxonomic microbiology studies, 16S ribosomal RNA (rRNA) sequencing is a common amplicon sequencing method used to identify and compare bacteria present within a given sample or from sample derived from complex microbiomes or environment that are difficult or impossible to study. Data obtained from this study are used to improve the sensitivity and specificity of taxonomic assignments down to the species level.

Molecular Genetics Of Microbes

Microbial whole genome sequencing is an important tool for mapping genomes of novel organism, completing genomes of known organisms or comparing genomes across multiple samples. Sequencing the entire genome is important for generating accurate reference genomes for identification and other comparative genomic studies in future.

Environmental Studies

Shotgun metagenomic sequencing allows researchers to comprehensively sample all genes in all organism present in a given complex sample. The method enables microbiologist to evaluate bacterial diversity and detect the abundance of microbes in various environments. Shotgun metagenomics also provide a means to study unculturable microorganisms that are otherwise difficult or impossible to analyse.

New and emerging diseases on metagenomics has become an interestingly important tool in helping to better understand human health, current and new emerging diseases, environmental system and the overall taxonomic classification of microbes.

Medicine

Medical technicians can also use sequencing technology to sequence genes from patients to determine if there is any risk of them having genetic diseases. This is a form of genetic testing, which may not involve DNA sequencing solely.

Forensics

For forensic identification and paternity testing, DNA sequencing are used along with DNA profiling methods.

TYPES OF SEQUENCNG AND APPLICATIONS

DNA Sequencing

DNA sequencing is a technology that is applied to determine the order of nucleotides of the four bases adenine (A), guanine (G), cytosine (C) and thymine (T) in a strand of DNA. Information of DNA sequences has become important for basic biological research and in numerous applied fields such as in plant breeding, forensic biology, medical diagnosis, biotechnology, biological systematics and virology. The rapid speed of sequencing achieved with modern DNA sequencing technology has been applied in the sequencing of complete DNA sequences genomes of various types of species such as animal, plant and microbial species.

Maxam-Gilbert Sequencing

Maxam-Gilbert sequencing uses chemical modification on DNA and cleavage at specific bases. This method allowed to use purified samples of double-stranded DNA without further cloning. However, this method's used radioactive labelling and its technical procedure is extremely complex.

This sequencing method requires radioactive labelling at 5' end of the fragmented purified DNA strand. Through chemical treatment, breaks at a small proportion of one or two of the four nucleotide bases in each of four reactions (G, A+G, C, C+T) are initiated. However, the concentration of the modifying chemicals are controlled to introduce one modification per DNA molecule.

Thus, a series of labelled fragments are generated from the radiolabelled end to the first breakage site in each molecule. The fragments in the four reactions are electrophoresed parallelly in denaturing acrylamide gels for size separation. To visualize the separated fragments, the gel is exposed to X-ray film for autoradiography. Each dark bands corresponding to a radiolabelled DNA fragment, from which the sequence may be inferred.

Chain-Termination Methods

The chain-termination method developed by Frederick Sanger and co-workers in 1977 had become the method of choice. It is because of its relative ease and reliability. When invented, the chain-terminator method used fewer toxic chemicals and low amounts of radioactivity than the Maxam and Gilbert method. Because of its comparative ease and low in radioactivity exposure, the Sanger method was soon become the method of choice used in the first generation of DNA sequencers. Sanger sequencing developed from fluorescent labelling, capillary electrophoresis, to general automation. These developments allowed much more efficient sequencing with lower costs. The Sanger method technology was successfully used to produce the first human genome in 2001. Large-scale sequencing can be achieved through sequencing of very long DNA pieces or by generating very large numbers of short sequences.

For longer targets such as chromosomes, restriction enzymes or shearing with mechanical forces, large DNA fragments into shorter DNA fragments will be produced. The fragmented DNA may then be cloned into a DNA vector and amplified in a bacterial host such as *Escherichia coli*. Short DNA fragments purified from bacterial colonies are individually sequenced and assembled electronically into one long, contiguous sequence by software. The size selection step to collect DNA fragments of uniform size can improve sequencing efficiency and accuracy of the genome assembly.

De novo sequencing

De novo sequencing used to determine the sequence of DNA from an unknown sequence. Gaps in the assembled sequence will be filled by primer walking. Most sequencing approaches use an *in vitro* cloning step to amplify individual DNA molecules, because their molecular detection methods are not sensitive enough for a single molecule sequencing.

Shotgun Sequencing

Shotgun sequencing are often used for sequencing large genomes, but its assembly is complex and difficult due to sequence repeats that cause gaps in genome assembly. It is designed for analysis of DNA sequences longer than 1000 base pairs or including entire chromosomes. This method requires the target DNA to be broken into random fragments. After sequencing individual fragments, the sequences can be reassembled on the basis of their overlapping regions.

Bridge PCR

Bridge PCR is another method for *in vitro* clonal amplification in which DNA fragments are amplified upon primers attached to a solid surface and form DNA colonies or DNA clusters. This method is used in the Illumina Genome Analyzer sequencers. In single-molecule methods, they use bright fluorophores and laser excitation to detect base addition from individual DNA molecules fixed to a surface and eliminating the need for molecular amplification.

High-throughput formerly next-generation sequencing applies to genome sequencing, genome resequencing, transcriptome profiling (RNA-Seq), DNA-protein interactions (ChIP-sequencing) and epigenome characterization. Resequencing is necessary, because the genome of a single individual of a species will not indicate all of the genome variations among other individuals of the same species.

The high demand for low-cost sequencing has driven the development of high-throughput sequencing technologies that fastened the sequencing process by producing thousands or millions of sequences concurrently.

High-throughput sequencing technologies are intended to lower the cost of DNA sequencing beyond what is possible with standard dye-terminator methods. In ultra-high-throughput sequencing as many as 500,000 sequencing-by-synthesis operations can run simultaneously.

Massively Parallel Signature Sequencing (MPSS)

The massively parallel signature sequencing (MPSS) are the first high-throughput sequencing technologies. MPSS was based on bead method that used a complex approach of adapter ligation followed by adapter decoding and reading the sequence in increments of four nucleotides. This method made it susceptible to sequence-specific bias or loss of specific sequences. Since MPSS is a complex technology it is not sold to independent laboratories. However, the essential properties of the MPSS output were include hundreds of thousands of short DNA sequences. In the case of MPSS, these were typically used for sequencing cDNA for measurements of gene expression levels.

Polony Sequencing

The polony sequencing method was among the first high-throughput sequencing systems that was used to sequence a full *E. coli* genome. It combined an in vitro paired-tag library with emulsion PCR, an automated microscope and ligation-based sequencing chemistry to sequence an *E. coli* genome at an accuracy of >99.99%.

454 Pyrosequencing

Pyrosequencing was developed by 454 Life Sciences and until today was acquired by Roche Diagnostics. The method amplifies DNA inside emulsion PCR which consist of water droplets in an oil solution. Each droplet containing a single DNA template attached to a single primer-coated bead that then forms a clonal colony. The sequencing machine contains many wells in picoliter-volume each containing a single bead and the sequencing enzymes. Pyrosequencing uses luciferase to generate light for detecting the individual nucleotides added to the nascent DNA. The combined data are then used to generate sequence read-outs.

Illumina (Solexa) Sequencingsolex A Is now part of Illumina and developing a sequencing method based on reversible dye-terminators technology and engineered polymerases. It is based on the clonal amplification of DNA on a surface. In this method, DNA molecules and primers are first attached on a slide or flow cell and amplified with polymerase so that local clonal DNA colonies later become DNA clusters. To determine the sequence of study, four types of reversible terminator bases are added and non-incorporated nucleotides are washed away. A camera is connected to takes images of the fluorescently labelled nucleotides. Then the dye, along with the terminal 3' blocker, is chemically removed from the DNA, allowing for the next cycle to take over. Unlike pyrosequencing, the DNA chains are extended one nucleotide at a time and image acquisition will be performed at a delayed moment, allowing for very large arrays of DNA colonies to be captured by sequential images taken from a single camera. Decoupling the enzymatic reaction and the image capture allows for optimal throughput and theoretically unlimited sequencing capacity.

SOLiD Sequencing

SOLiD sequencing technology employs sequencing by ligation. A pool of oligonucleotides of a fixed length are labelled according to the sequenced position. Oligonucleotides are annealed and ligated to a preferential ligation by DNA ligase to matched the sequences results in a signal informative of the nucleotide at a particular position. Before sequencing, the DNA is amplified by emulsion PCR. The resulting beads, each containing single copies of the same DNA molecule, are deposited on a glass slide. The result consists of number of a sequences and lengths which comparable to Illumina sequencing.

Ion Torrent Semiconductor Sequencing

Ion Torrent semiconductor sequencing was developed by a system based on standard sequencing chemistry with a semiconductor as a detection system. This method of sequencing is based on the detection of hydrogen ions that are released during the polymerisation of DNA. A microwell containing a template DNA strand to be sequenced is flooded with a single type of nucleotide. If the introduced nucleotide is complementary to the leading template nucleotide, it will incorporated into the growing complementary strand. This causes the release of a hydrogen ion and triggers the hypersensitive ion sensor, which indicates that a reaction has occurred. If homopolymer repeats are present in the template sequence, multiple nucleotides will be incorporated in a single cycle. This leads to a corresponding number of released hydrogens and a proportionally higher electronic signal.

DNA Nanoball Sequencing

DNA nanoball sequencing is a type of high throughput sequencing technology used to determine the entire genomic sequence of an organism. The method uses rolling circle replication to amplify small fragments of genomic DNA into DNA nanoballs. Unchained sequencing by ligation is then used to determine the nucleotide sequence. This method of DNA sequencing allows large numbers of DNA nanoballs to be sequenced per run at a low reagent costs compared to other high-throughput sequencing platforms. However, only short sequences of DNA are determined from each DNA nanoball and mapping is done by using the short reads to a reference genome which is usually a difficult task.

Heliscope Single Molecule Sequencing

Heliscope sequencing is a method of single-molecule sequencing developed by Helicos Biosciences. It uses DNA fragments with added poly-A tail adapters which are then attached to the flow cell surface. The next steps involve with extension-based sequencing with cyclic washes of the flow cell with fluorescently labelled nucleotides. The reads which is short in length 35 bp are performed by the Heliscope sequencer.

Single Molecule Real Time (SMRT) Sequencing

SMRT sequencing is also referred to as a third-generation or long-read sequencing. SMRT sequencing is based on synthesis approach sequencing. The DNA is synthesized in a zero-mode wave-guides (ZMWs) a small well-like containers with the capturing tools located at the bottom of the well. The sequencing is performed with the use of unmodified polymerase attached to the ZMW bottom and fluorescently labelled nucleotides flowing freely in the solution. The wells are constructed in a way that only the fluorescence occurring by the bottom of the well is detected. The fluorescent label is detached from the nucleotide upon its incorporation into the DNA strand leaving an unmodified DNA strand. This methodology allows detection of nucleotide modifications such as cytosine methylation. This happens through the observation of polymerase kinetics. This approach allows reads of 20,000 nucleotides or more, with average read lengths of 5 kilobases.

Nanopore DNA Sequencing

Nanopore sequencing is referred to as "third-generation" or "long-read" sequencing, along with SMRT sequencing. In nanopore DNA sequencing the DNA passing through the nanopore changes its ion current. This change is dependent on the shape, size and length of the DNA sequence. Each type of the nucleotide blocks the ion flow through the pore on a different period of time. The method does not require modified nucleotides and is performed in real time.

This concept originated from the idea that single stranded DNA or RNA molecules can be electrophoretically driven in a strict linear sequence through a biological pore that can be less than eight nanometers and can be detected by the release of an ionic current while moving through the pore. The pore contains a detection region capable of recognizing different bases, with each base generating various time specific signals corresponding to the sequence of bases as they cross the pore. Precise control over the DNA transport through the pore is crucial for success. Various enzymes such as exonucleases and polymerases have been used to moderate this process by positioning them near the pore's entrance.

Two main areas of nanopore sequencing in development are solid state nanopore sequencing and protein based nanopore sequencing. Protein nanopore sequencing utilizes membrane protein complexes such as ∝-Hemolysin, MspA (Mycobacterium Smegmatis Porin A) or CssG, which show a great promise on their ability to distinguish between individual and groups of nucleotides. In contrast, solid-state nanopore sequencing utilizes synthetic materials such as silicon nitride and aluminium oxide and it is preferred for its superior mechanical, thermal and chemical stability. The fabrication method is essential for this type of sequencing as the nanopore array can contain hundreds of pores with diameters smaller than eight nanometers.

Tunnelling Currents DNA Sequencing

Another approach for DNA sequencing is through measurements of the electrical tunnelling currents across single-strand DNA as it moves through a channel. Depending on its electronic structure, each base affects the tunnelling current differently, allowing differentiation between different bases.

The use of tunnelling currents has the potential to sequence the orders of magnitude faster than ionic current methods and the sequencing of several DNA oligomers and micro-RNA has already been achieved so far.

Sequencing by Hybridization

Sequencing by hybridization is a non-enzymatic method that uses DNA microarray techniques. A single pool of DNA from specific sequence will be detected once it binds to fluorescently labelled and hybridized to a known sequence. Strong hybridization signals from a given spot on the array identifies the target sequence in the sequenced DNA.

This method of sequencing utilizes binding characteristics of a short single stranded DNA library molecules similar to DNA probes. Non-specific hybrids are removed by washing and the target DNA in a hybrid forms is eluted. Hybrids are re-arranged such that the DNA sequence can be reconstructed again. The advantage of this sequencing is its ability to capture a large number of targets with a homogenous coverage. In this type of sequencing a large number of chemicals and starting DNA is usually required. However, with the advent of solution-based hybridization, much less equipment and chemicals are required.

Sequencing With Mas Spectrometry

Basically, mass spectrometry can be used to determine DNA sequences. MALDI-TOF MS, has specifically been investigated as an alternative method of gel electrophoresis for visualizing DNA fragments. With this method, DNA fragments generated by chain-termination sequencing reactions can be compared by mass rather than by size. It is because the mass of each nucleotide is different and this difference is only detectable by mass spectrometry. Moreover, single-nucleotide mutations in a fragment can also be more easily detected with MS than by gel electrophoresis. Furthermore, MALDI-TOF MS can more easily detect differences between RNA fragments, so researchers may indirectly sequence DNA with MS-based methods by first converting it to RNA.

The higher resolution of DNA fragments allowed by MS-based methods eased researchers in forensic science to identify individuals by determining single-nucleotide polymorphisms in human DNA samples from mitochondrial DNA for its higher stability and applications for lineage studies. However, MS-based DNA sequencing may not be suitable for large *de novo* sequencing projects.

Microfluidic Sanger Sequencing

In microfluidic Sanger sequencing the complete thermocycling amplification of DNA fragments as well as their separation by electrophoresis is done on a single glass wafer (approximately 10 cm in diameter) thus reducing the reagent usage as well as cost. In some circumstances researchers have shown that they can increase the throughput of conventional sequencing through the use of microchips.

Microscopy-Based Techniques

This approach directly visualizes the sequence of DNA molecules using electron microscopy.

RNAP Sequencing

This method is based on the usage of RNA polymerase (RNAP), which is attached to a polystyrene bead. One end of DNA to be sequenced is attached to another bead, with both beads being placed in optical traps. RNAP motion during transcription brings the beads in closer and their relative distance changes, this are then will be recorded as a single nucleotide resolution. The sequence is deduced based on the four readouts with lowered concentrations of each of the four nucleotide types. A comparison is made between regions and sequence information by comparing the known sequence regions to the unknown sequence regions.

In Vitro Virus High-Throughput Sequencing

A method has been developed to analyze full sets of protein interactions using a combination of 454 pyrosequencing and an *in vitro* virus mRNA display method. Specifically, this method covalently links proteins of interest to the mRNAs that encoding them, then detects the mRNA pieces using reverse transcription PCRs. The mRNA may then be amplified and sequenced. The combined method was titled as IVV-HiTSeq and can be performed under cell-free conditions, though its results may not be representative of *in vivo* conditions.

Sample Preparation

The success of a DNA sequencing protocol is dependent on the sample preparation. A successful DNA extraction will yield a sample with long, non-degraded strands of DNA which require further preparation according to the sequencing technology that will be used. For Sanger sequencing, either cloning procedures or PCR are required prior to sequencing. In the case of next generation sequencing methods, library preparation is required before processing. With the advent of next generation sequencing, Illumina and Roche 454 methods have become a common approach for transcriptomic studies (RNAseq). RNA can be extracted from tissues of interest and converted to complementary DNA (cDNA) using reverse transcriptase (a DNA polymerase that synthesizes a complimentary DNA based on existing strands of RNA in a PCR-like manner). Complementary DNA can be processed the same way as genomic DNA, allowing the expression levels of RNAs to be determined for the selected tissue.

Gene knockout

From the name itself a gene knockout is a genetic technique in which one of an organism genes is knocked out of the organism. Also known as knockout organisms or simply knockouts, they are used in learning about a gene that has been sequenced, but which has an unknown or incompletely known function. Researchers draw inferences from the difference between the knockout organism and normal individuals.

The technique is essentially the opposite of a gene knocking. Knocking out two genes simultaneously in an organism is known as a double knockout (DKO). Similarly the terms triple knockout (TKO) and quadruple knockouts (QKO) are used to describe the number of genes that will be knocked out three or four knocked out genes, respectively.

Knockout is accomplished through a combination of techniques, beginning in the test tube with a plasmid, a bacterial artificial chromosome or other DNA construct and proceeding to cell culture. Individual cells are genetically transfected with the DNA construct. Often the goal is to create a transgenic animal that has the altered gene. If so, embryonic stem cells are genetically transformed and inserted into early embryos. Resulting animals with the genetic change in their germline cells can then often pass the gene knockout to future generations.

To create knockout moss, transfection of protoplasts is the preferred method. Such transformed *Physcomitrella*-protoplasts directly regenerate into fertile moss plants. Eight weeks after transfection, the plants can be screened for gene targeting via PCR. The construct is engineered in a way to recombine it with the target gene, which is carried out by incorporating sequences from the gene itself into the construct. Recombination then occurs in the region of that sequence within the gene, resulting in the insertion of a foreign sequence to disrupt the whole gene. With its sequence interrupted, the altered gene in most cases will be translated into a nonfunctional protein.

A conditional knockout allows gene deletion in a tissue on the basis of time specific manner. This is done by introducing short sequences called loxP sites around the gene. These sequences will be introduced into the germ-line via the same mechanism as a knock-out. This germ-line can then be crossed to another germline containing Cre-recombinase which is a viral enzyme that can recognize these sequences, recombines them and deletes any genes that flanked by these sites.

Because of the desired type of DNA recombination is a rare event in the case of most cells and most constructs, the foreign sequence chosen for insertion usually includes a reporter. This enables easy selection of cells or individuals in which knockout will be successful. Sometimes the DNA construct inserts into a chromosome without the desired homologous recombination with the target gene. To eliminate such cells, the DNA construct often contains a second region of DNA that allows such cells to be identified and discarded.

In diploid organisms, which contain two alleles for most genes and may as well contain several related genes that collaborate in the same role, additional rounds of transformation and selection are performed until every targeted gene is knocked out. Selective breeding may be required to produce homozygous knockout animals.

Applications

Knockouts are primarily used to understand the role of a specific gene or DNA region by comparing the knockout organism to a wildtype with a similar genetic background.

Knockout organisms are also used as screening tools

- to developed drugs
- to target specific biological processes
- to identified deficiencies in biological process
- to understand the mechanism of action of a drug

Exome Sequencing

Exome sequencing also known as whole exome sequencing (WES). It is a transcriptomics technique for sequencing all or any expressed genes in a genome (known as the exome). It consists of two steps:

- The first step is to select only the subset of DNA that encodes proteins. These regions are known as exons for example humans have about 180,000 exons, constituting about 1% of the human genome, or approximately 30 million base pairs.

- The second step is to sequence the exonic DNA using any high-throughput DNA sequencing technology.

The goal of this approach is to identify genetic variants that alter protein sequences and through this technique one can do this at a much lower cost than whole-genome sequencing. Since these variants can be responsible for both Mendelian and common polygenic diseases, such as Alzheimer's disease, whole exome sequencing has been applied both in academic research and as a clinical diagnostic.

Key strength:

- Identifies variants across a wide range of applications, including populations genetics, genetic disease and cancer studies
- Achieves comprehensive coverage of expertly selected exonic content
- Provides a cost effective alternative to whole genome sequencing, as the human genome contains~85% of known disease causing variants

Cancer research

Exome sequencing allows researchers to focus on only the coding regions. By sequencing only 1-2% of the genome it is possible to sequence to higher depth more economically. By performing tumor-normal exome sequencing, researchers can focus in on the changes most likely to be tumorigenic.

Complex disease

Genetic association and linkage studies with exome sequencing provide researchers with insights into complex diseases through common genetic variants shared between large numbers of individual. In recent years, researchers have begun to extend complex trait association studies to focus on rare, coding variants that often have impactful functional consequences on affected patients. Researchers taking advantage of the affordability of exome sequencing gain the genomic tools to discover and test for associations between rare coding variants and complex disease directly testing the rare variant common disease hypothesis.

Clinical research

Exome sequencing is an increasingly powerful tool for investigation human disease. It allows evaluation of thousands of protein-coding regions simultaneously with a single test. Exome sequencing accelerates discovery through focused analysis. In the past, clinical genetic tests were chosen based on the clinical results of the patient either focused on one gene or a small number of known genes to be associated with a particular syndrome, or surveyed only certain types of variation through comparative genomic hybridization. This has provided definitive genetic diagnoses in fewer than half of all patients that was tested. Exome sequencing is now increasingly used to find mutations in genes which already known to cause disease as well as to identify novel genes by comparing exomes from patients with similar features.

Mendelian disorder discovery

Mendelian disorder discovery can be carried out through exome sequencing. A few unrelated individuals were sequenced through this technique as it is a cost effective for identifying the genes related to rare Mendelian disorders and unsolved Mendelian disorders in all of an individual's genes. These diseases are most often caused by very rare genetic variants that are only present in a tiny number of individuals. Furthermore, because severe disease-causing variants are much more likely (but by no means exclusively) to be in the protein coding sequence, focusing on this 1% costs are far less than whole genome sequencing but still capable to detects a high yield of relevant variants.

Methodology

Step 1: Target-enrichment strategies

Target-enrichment methods allow the users to selectively capture genomic regions of interest from a pool of DNA sample prior to sequencing. Microarrays contain single-stranded oligonucleotides with sequences from the human genome to tile the region of interest fixed to the surface. Genomic DNA is sheared to form double-stranded fragments. The fragments undergo end-repair to produce blunt ends and adaptors with added universal priming sequences. These fragments are hybridized to oligos on the microarray. Non-hybridized fragments are washed away and the desired fragments are eluted. The fragments are then amplified using PCR. This method is both time-saving and cost-effective compared to PCR based methods. Limitations in this technique include the need for expensive hardware as well as a relatively large amount of DNA.

In-Solution Capture

Therefore, to capture genomic regions of interest using in-solution capture, a pool of custom oligonucleotides probes is synthesized and hybridized in a solution of a fragmented genomic DNA sample. The probes which is labeled with beads will selectively hybridize to the genomic regions of interest. The beads including the DNA fragments of interest will be pulled down and washed to clear excess material. The beads are then removed and the collected genomic fragments will be sequenced.

This method was developed to improve the hybridization capture target-enrichment method. In-solution capture as opposed to hybrid capture, there is an excess of probes to target regions of interest over the amount of template required. The optimal target size is about 3.5 megabases and yields excellent sequence coverage of the target regions. The preferred method is dependent on several factors including number of base pairs in the region of interest, demands for reads on target and equipment in-use.

Step 2: Sequencing

There are several sequencing platforms besides the classical Sanger sequencing such as the Roche 454 sequencer, the Illumina Genome Analyzer II and the Life Technologies SOLiD & Ion Torrent all of which have been used for exome sequencing.

Key strengths

- Combine short inserts and longer reads to allow characterization of any genome
- Provides high quality, long contig assemblies for de novo whole genome sequencing'

- Reveals disease-causing alleles that might not have been identified otherwise
- Captures results for new variants without requiring resequencing

Cancer Research

In cancer research, whole genome sequencing can be employed to monitor genomic changes that occur in a tumor, including the presence of markers for good and poor prognosis. Through tumor-normal whole genome sequencing, researchers can identify the tumor genome and determine differences from non-tumor DNA.

Complex Disease

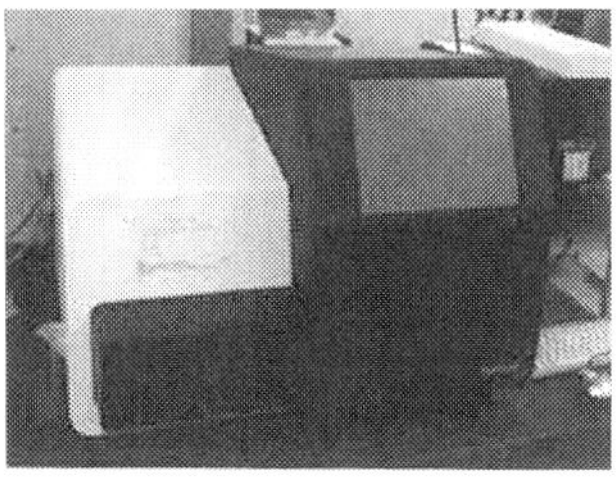

Next generation sequencing (NGS) is proving to be a powerful tool for genome wide association studies (GWAS). GWAS allows researchers to identify common genetic variants across populations to establish links between those variants and the traits of individual, including predisposition to disease.

Genetic disease

Phased sequencing complements human whole genome sequencing by distinguishing between the genetic contributions from each parent to the chromosomes of the offspring. Inheritance patterns are often includes the understanding of the basis of genetic disease.

Agrigenomics

Exponential population growth and a changing climate are creating unique challenges for people working to maintain the food supply. To overcome food supply challenges, many are turning into science. NGS are particularly useful in agriculture research, where genomes can be complex and prior knowledge scarce.

STEP 1: CHOOSE LIBARARY PREP

STEP 2: CHOOSE SEQUENCER

STEP 3: OBTAIN RESULTS; ANNOTATE/FILTER

Whole Genome Sequencing (small genomes)

Key strengths:

- Allows investigation of all genes in all organisms present in a given complex sample
- Sequences thousands of organisms in parallel
- Sequences and discovers unculturable organisms
- Detects very low abundance members of the microbial community that might be missed or are too expansive to identify using other methods
- Provides comprehensive analysis of the microbial/viral genome
- Aids discovery of new biomarkers (SNP variants) within a microbial/viral sample by providing distinct gene information from homologous chromosomes, supporting haplotyping and allele-specific solutions

Microbiology

Whole genome sequencing is an important tool for mapping genomes of novel organisms, finishing genomes of known organisms, comparing genomes across multiple samples and enables the study of pathogen evolution. Sequencing the entire microbial genome is important for generating accurate reference genomes, for microbial identification and other comparative genomic studies.

Microbial outbreak identification

The power and speed of NGS was demonstrated during the 2011 entero-aggregative *E. coli* outbreak in Europe, which prompted a rapid scientific response. Using NGS data, researchers quickly generated a high quality whole genome sequence of the bacterial strain, enabling them to increase our understanding of the genetic mutations conferring the increased virulence.

Bacterial and viral screening/identification

Microbiologist and clinicians can use WGS for small genomes towards the identification and confirmation (screening purposes) bacterial genome in food testing and human/veterinary (agricultural testing) screening. The ability to effectively sequence a bacterial/viral genome provides the quick and accurate identification of potential pathogens, increasing true positives and decreasing false positive test results.

Bacterial Mutagenesis

High resolution genome data can be instrumental for examining pathogenesis, horizontal gene transfer, pan-genomes and co-evolution of hosts and symbionts/parasites. The wealth of information enabled by NGS is beneficial for mutational studies of all kinds, including directed evolution strategies, lab adaptation analyses, mutagenesis screens, or temporal and spatial dynamics of epidemics and transmission study.

Whole Genome Sequencing (Long-Range)

Key strength

- Complements whole genome sequencing by providing distinct gene information from homologous chromosomes, supporting haplotyping and allele-specific solutions
- Synthetically assembles long reads for de novo and genome finishing applications with high accuracy
- Offers the industry`s lowest DNA input requirement to prepare mata pair libraries as large as 12 kb
- Aids identification of structural variants in complex genomes

Cancer genetics

Longer sequence reads enable detection of difficult structural variant rearrangements (deletions, inversions, transversions) commonly found in cancer and genetic disease.

Genetic Disease

Longer sequence reads allow for phasing studies that capture unique chromosomal content, including mutations that might differ across chromosome copies. This technique allows researchers to distinguish between maternally and paternally inherited alleles, a distinction that helps establish the association between the number and type of gene mutations and the presence of disease.

Agrigenomics

Longer sequence reads allow researchers to produce more accurate genome assemblies in complex genomes as found in agricultural plants.

Molecular genetics

Long read de-novo sequencing is an important tool for mapping genomes of novel organisms, completing genomes of known organisms, or comparing genomes across multiple samples. Sequencing the entire genome is important for generating accurate reference genomes.

Epigenetics

Key strength

- Enables investigation of epigenetic impact on gene regulation
- Enhances characterization of phenotype of interest
- Identifies potential biomarkers for a range of phenotypes
- Reveals links between epigenetic change and disease development

Cancer Research

Aberrant methylation is a common epigenetic change in cancer. Sequencing based methylation analysis enables researchers to identify and tract cancer methylation patterns by directly sequencing the methylated genome.

Complex disease

Genetics variants often have functional consequences that contribute to disease. These variants might be inherited or be the result of disease or exposure to environmental mutagen. NGS allows researchers to clarify the relationship between disease and the genetic variants.

Agrigenomics

Changes to DNA methylation are being investigated for potential down regulation of stress genes during drought, salinity changes, heat, cold and other environmental changes.

Imprinting

Environmental exposures are able to alter the methylation patterns of the genome along with other epigenetic changes. NGS methylation studies allow researchers a more complete understanding of the etiology of complex genetic diseases, clarifying the contribution of environment and genetic variation.

Amplicon sequencing

Key Strength

- Offers unparalleled efficiency for discovering, validating and screening genetic variants with highly targeted approach.
- Multiplexes up to 1536 amplicons per reaction to achieve coverage of up to 650 kb of cumulative sequence with TruSeq Custom Amplicon Kit
- Delivers improved performance in difficult to address GC-rich regions, as well as increased library yield, uniformity and stability from samples as low as 50 ng with TruSeq Custom Amplicon v1.5

Cancer Research

Custom amplicon sequencing is useful for targeting genes known to be involved in diseases such as cancer. Efficient, targeted custom amplicon sequencing achieves research results quickly and efficiently.

Microbial genomics

Microbial samples are often complex, requiring researchers to identify and compare the populations and activities of the microbes present within single sample. High speed, multiplex 16S amplicon sequencing improves the sensitivity and specificity of taxonomic assignments down to the genus level and species level in some cases.

Public health food borne illness occurs when people eat food contaminated with pathogenic organisms. NGS allows population health experts to quickly identify the organism involved in an outbreak, improving health care for the affected people and effectively addressing the source of the outbreak.

Cancer research sequencing

Key strength

- Sequences key genes or regions of interest to high depth using predesigned, analytically validated panels
- Provides cost effective, actionable findings for cancer- relevant genes
- Delivers easy to interpret results
- Obtains highly accurate results with variants present down to 5% variant allele frequency

Germline testing

The current paradigm is to test individuals believed to be the risk for a few select genes, providing a limited picture and potentially unclear results. NGS enables multiple risk associated genes to be examined in a single test, providing an expanded view rapidly, accurately and economically

Somatic variant detection

Iterative tests on cancer samples delay time to answer and can result in exhaustion of limited tissue. NGS enables multiple genes to be assessed in a single test, providing a single streamlined workflow, decreasing time to answer and producing accurate and highly sensitive results.

Disease and condition sequencing panels

Key Strength

- Detects rare variants that might be associated with genetic diseases and enables confident identification of causative de novo or inherited rare mutations in a single test.
- Provides cost effective, streamlined, targeted NGS of specific genetic diseases or conditions
- One sequencing panel, thousands of inherited conditions

The TruSightOne Sequencing Panel is a solution for germline mutation screening of exonic regions in 4,813 genes relevant for clinical research. Designed to cover the most commonly ordered molecular assays, the panel enables labs to perform these assays on site with single solution. At the same time, TruSight One facilitates the creation of several smaller sequencing panels with a common workflow, reducing both operational overhead and processing frequency. Genomic targets were identified based on information in the Human Gene Mutation Database (HGMD Professional), the Online Mendelian Inheritance in Man (OMIM) catalog, GeneTests Org, Illumina TruSight panels and other commercially available sequencing panels.

Inherited cardiac disease research

The increased information obtained from NGS is providing remarkable insight into the genomic and environmental components that underline inherited cardiac conditions. TruSight Cardio is designed to affordably and accurately screen 174 genes related to 20 inherited cardiac conditions to provide the greatest chance of revealing causative variants.

Inherited disease

TruSight Inherited Disease was initially based on a panel designed for preconception carrier testing for severe, recessive childhood diseases. It addresses the needs of medical geneticists with a primary focus on severe recessive diseases with childhood onset.

Autism

Specific genes are emerging as central to the understanding of autism.

These genes include genes that have been reported in the Online Mendelian Inheritance in Man (OMIM) database on autism, genes with recurrent mutations resulting in developmental delays, genes with reported mutations as found in case studies involving developmental delay characteristics and genes from summaries of autism-relevant genes (e. AutismKB4)

Applications of Exome Sequencing

Rare variant mapping in complex disorder

Current association studies have focused on common variation across the genome, as these are the easiest methods to apply. However, disease-causing variants of large effect have been found to lie within exomes in candidate gene studies and because of negative selection they are found in much lower allele frequencies and may remain unidentified in current standard genotyping assays. Believe, whole genome sequencing is a potential method to assay novel variant across the genome. However, in complex disorders, a large number of genes are thought to be associated with single disease risk. This underlying risk of heterogeneity means that large sample sizes are required for gene discovery and thus whole genome sequencing is not particularly cost-effective. In addition, variants in coding regions have been much more extensively studied and their functional implications are much more easier to be determined, making the practical applications of variants within the targeted exome region are more immediately accessible.

Discovery of Mendelian Disorders

In Mendelian disorders one or a very small number of variants within coding genes represent the entire condition. Because of the severity of these disorders, a few causal variants are presumed to be extremely rare or novel in the populationand would be missed out by any standard genotyping assay. Exome sequencing provides high coverage variant calls across coding regions, which are needed to separate true variants from noise. A successful model of Mendelian gene discovery involves the discovery of the de novo variants using trio sequencing, where parents and proband was genotyped.

Clinical Diagnostics

Exome sequencing can be used to diagnose the genetic disease in a patient. Identification of the underlying disease gene mutation(s) can have major implications for diagnostic and therapeutic approaches, prediction of disease history and makes it possible to test at-risk family members. There are many factors that make exome sequencing superior to single gene analysis:

- the ability to identify mutations in genes that were not tested due to a typical clinical presentation

  or

- the ability to identify clinical cases where mutations from different genes contribute to the different phenotypes in the same patient

Having diagnosed a genetic cause of a disease, this information may guide the selection of appropriate treatment.

Case study 1: Infant with inflammatory bowel disease. Conventional diagnostics had previously been used, but the results could not explain the infant's symptoms. Analysis of exome sequencing data identified a mutation in the XIAP gene. Knowledge of this gene's function guided the infant's treatment, leading to a bone marrow transplantation which cured the child from the disease.

Case study 2: Patient with Bartter Syndrome and congenital chloride diarrhea Bilgular's. Used exome sequencing and identified the underlying mutation for a patient with severe brain malformations.

Exome sequencing will be particularly valuable for gene discovery in those conditions in which mapping has been confounded by locus heterogeneity and uncertainty about the boundaries of diagnostic classification and its broad application in medicine.

ChIP Sequencing

ChIP is a powerful method to selectively enrich the DNA sequences bound by a particular protein in the living cells. ChIP-seq is used primarily to determine the phenotype-affecting mechanisms mediated by transcription factors and other chromatin-associated proteins. Determining the regulation of proteins interact with DNA is essential for fully understand the biological processes and disease states. This epigenetic information is complementary to genotype and expression analysis. ChIP-seq technology is currently seen primarily as an alternative to ChIP-chip sequencing which requires a hybridization array that can introduces some bias due to a fixed number of probes. ChIP Sequencing is thought to have less bias, although the sequencing bias of different sequencing technologies is not yet fully studied.

Specific DNA sites in direct physical interaction with transcription factors and other proteins can be isolated by chromatin immunoprecipitation. ChIP produces a library of target DNA sites bound to a protein of interest *in vivo*. This can be applied to the set of ChIP-able proteins and modifications, such as transcription factors, polymerases and transcriptional machinery, structural proteins and DNA modifications. As an alternative to the dependence on specific antibodies, different methods have been developed to find the superset of all nucleosome-depleted or nucleosome-disrupted active regulatory regions in the genome, like DNase-Seq and FAIRE-Seq. The ChIP process enriches specific crosslinked DNA-protein complexes using an antibody against the protein of interest. After size selection, all the resulting ChIP-DNA fragments are sequenced simultaneously using a genome sequencer. There are many new sequencing methods used in this sequencing step. Some technologies that analyze the sequences can use cluster amplification of adapter-ligated ChIP DNA fragments on a solid flow cell substrate to create clusters of approximately 1000 clonal copies each. The resulting high density array of template clusters on the flow cell surface is sequenced by a genome analyzing program. Each template cluster undergoes sequencing-by-synthesis in parallel using novel fluorescently labelled reversible terminator nucleotides. Templates are sequenced base-by-base during each read.

Sensitivity

- the depth of the sequencing run (i.e. the number of mapped sequence tags), the size of the genome and the distribution of the target factor. The sequencing depth is directly correlated with cost. If abundant binders in large genomes have to be mapped with high sensitivity, costs are high as an enormously high number of sequence tags will be required.
- the precision is not limited by the spacing of predetermined

probes. By integrating a large number of short reads, highly precise binding site localization is obtained.

- ChIP-seq data can be used to locate the binding site within few tens of base pairs of the actual protein binding site. Tag densities at the binding sites are a good indicator of protein–DNA binding affinity, which makes it easier to quantify and compare binding affinities of a protein to different DNA sites.

Analysis

As with many high-throughput sequencing approaches, ChIP-seq generates extremely large data sets, for which appropriate computational analysis methods are required. To predict DNA-binding sites from ChIP-seq read count data, peak calling methods have been developed. The most popular peak calling method is MACS which empirically models the shift size of ChIP-Seq tags and uses it to improve the spatial resolution of predicted binding sites.

Another relevant computational problem is differential peak calling. Here its identifies significant differences in two ChIP-seq signals from distinct biological conditions. Differential peak callers segment in two ChIP-seq signals and identify differential peaks can be carried out using Hidden Markov Models. While, examples for two-stage differential peak callers identification can be analysed using software known as ChIPDiff and ODIN.

Some examples of success Case Studies:

- STAT1 DNA association: ChIP-seq was used to study STAT1 targets in HeLA S3 cells.
- Nucleosome Architecture of Promoters: Using ChIP-seq, it was determined that Yeast genes have a minimal nucleosome-free promoter region of 150bp in which RNA polymerase can initiate transcription.
- Transcription factor conservation: ChIP-seq was used to compare conservation of TFs in the forebrain and heart tissue in embryonic mice. It was identified and validated that the transcription enhancers for the heart are less conserved than those for the forebrain during the same developmental stage.
- Genome-wide ChIP-seq: The worm *C. elegans* sequencing was

completed through ChIP sequencing. 20% of the annotated candidate genes were assigned to transcription factors. It has 22 transcription factors. Several transcription factors were assigned to non-coding RNA regions and functioned to developmental or environmental variables, regulate genes that control other transcription factors, serve as both targets and regulators of other factors and demonstrating a network of regulation.

- Inferring regulatory network: ChIP-seq signal of Histone modification were shown to be more correlated with transcription factor motifs at promoters in comparison to RNA level. Hence author proposed that using histone modification ChIP-seq would provide more reliable inference of gene-regulatory networks in comparison to other methods based on expression.

ChIP-seq offers an alternative to ChIP-chip. STAT1 experimental ChIP-seq data have a high degree of similarity to results obtained by ChIP-chip for the same type of experiment, with >64% of peaks in shared genomic regions. Because the data are sequence reads, ChIP-seq offers a rapid analysis pipeline. As long as a high-quality genome sequence is available for read mapping and the genome doesn't have repetitive content that confuses the mapping process, the potential to detect mutations in binding-site sequences is detectable and will directly support any observed changes in protein binding and gene regulation.

Sanger Sequencing

Sanger sequencing is a method of DNA sequencing first commercialized by Applied Biosystems, based on the selective incorporation of chain-terminating dideoxynucleotides by DNA polymerase during in vitro DNA replication. It was the most widely used sequencing method for approximately 39 years. More recently, higher volume Sanger sequencing has been supplanted by Next-Gen sequencing methods, especially for large-scale and automated genome analyses. However, the Sanger method remains in wide use for smaller-scale projects, validation of Next-Gen results and for obtaining especially long contiguous DNA sequence reads which is > 500 nucleotides.

Method:

The classical chain-termination method requires

- single-stranded DNA template,
- a DNA primer,
- a DNA polymerase,
- normal deoxynucleosidetriphosphates (dNTPs),
- modified di-deoxynucleotidetriphosphates (ddNTPs), the latter of which terminate DNA strand elongation.

These chain-terminating nucleotides lack a 3'-OH group required for the formation of a phosphodiester bond between two nucleotides, causing DNA polymerase to cease extension of DNA when a modified ddNTP is incorporated. The ddNTPs may be radioactively or fluorescently labeled for detection in automated sequencing machines.

The DNA sample is divided into four separate sequencing reactions, containing all four of the standard deoxynucleotides (dATP, dGTP, dCTP and dTTP) and the DNA polymerase. To each reaction is added only one of the four dideoxynucleotides (ddATP, ddGTP, ddCTP, or ddTTP), while the other added nucleotides are ordinary ones. The dideoxynucleotide is added to be approximately 100-fold lower in concentration than the corresponding deoxynucleotide (e.g. 0.005mM ddATP : 0.5mM dATP) allowing for enough fragments to be produced while still transcribing the complete sequence. Putting it in a more sensible order, four separate reactions are needed in this process to test all four ddNTPs. Following rounds of template DNA extension from the bound primer, the resulting DNA fragments are heat denatured and separated by size using gel electrophoresis. In the original publication of 1977, the formation of base-paired loops of ssDNA was a cause of serious difficulty in resolving bands at some locations. This is frequently performed using a denaturing polyacrylamide-urea gel with each of the four reactions run in one of four individual lanes (lanes A, T, G, C). The DNA bands may then be visualized by autoradiography or UV light and the DNA sequence can be directly read off the X-ray film or gel image.

In the image, X-ray film will be exposed to the gel and the dark bands are correspond to DNA fragments of different lengths. A dark band in a lane indicates a DNA fragment that is the result of chain termination after incorporation of a dideoxynucleotides (ddATP, ddGTP, ddCTP, or ddTTP). The relative positions of the different bands among the four lanes, from bottom to top, are then used to read the DNA sequence.

Technical variations of chain-termination sequencing include tagging with nucleotides containing radioactive phosphorus for radio-labelling, or using a primer labeled at the 5' end with a fluorescent dye. Dye-primer sequencing facilitates reading in an optical system for faster and more economical analysis and automation. The later development by Leroy Hood and coworkers of fluorescently labeled ddNTPs and primers set the stage for automated, high-throughput DNA sequencing.

Chain-termination methods have greatly simplified DNA sequencing. For example, chain-termination-based kits are commercially available that contain the reagents needed for sequencing, pre-aliquoted and ready to use. Limitations include non-specific binding of the primer to the DNA, affecting accurate read-out of the DNA sequence and DNA secondary structures affecting the fidelity of the sequence.

Dye-terminator sequencing

Dye-terminator sequencing utilizes labelling of the chain terminator ddNTPs, which permits sequencing in a single reaction, rather than four reactions as in the labelled-primer method. In dye-terminator sequencing, each of the four dideoxynucleotide chain terminators is labelled with fluorescent dyes, each of which emit light at different wavelengths.

Owing to its greater expediency and speed, dye-terminator sequencing is now the mainstay in automated sequencing. Its limitations include dye effects due to differences in the incorporation of the dye-labelled chain terminators into the DNA fragment, resulting in unequal peak heights and shapes in the electronic DNA sequence trace chromatogram after capillary electrophoresis .

This problem has been addressed with the use of modified DNA polymerase enzyme systems and dyes that minimize incorporation variability, as well as methods for eliminating "dye blobs". The dye-terminator sequencing method, along with automated high-throughput DNA sequence analyzers, is now being used for the vast majority of sequencing projects.

Automation and sample preparation

Automated DNA-sequencing instruments are capable to sequence up to 384 DNA samples in a single batch runs. A standard batch runs can go up to 24 times a day. A DNA sequencers separate strands by size or length through capillary electrophoresis. DNA sequencers can also detect and record dye fluorescence and output data as fluorescent peak trace chromatograms. During sequencing reactions a numbers of step are performed separately before loading samples onto the sequencer, there are such as thermocycling process and labelling and cleanup and re-suspension of samples in a buffer solution. A number of commercial and non-commercial software packages that available commercially can be used to trim low-quality DNA traces automatically. These programs score the quality of each peak and remove low-quality base peaks which are generally located at the ends of the sequence.

Challenges

Common challenges of DNA sequencing with the Sanger method include poor quality sequencing in the first 15-40 bases of the sequence due to inefficient primer binding and deteriorating quality of sequencing traces after 700 bases. However, base calling software such as Phred will typically provide an estimation on quality to aid in trimming of low-quality regions of sequences.

In cases where DNA fragments are cloned before sequencing, the resulting sequence may contain some portion of the cloning vector. In contrast, PCR-based cloning and next-generation sequencing technologies based on pyrosequencing often avoid using cloning vectors. Recently, one-step Sanger sequencing such as Ampliseq and SeqSharp comes with combined amplification and sequencing methods. It was developed to allow rapid sequencing of target genes without cloning or prior amplification.

Current methods can directly sequence only relatively short nucleotides ranging from 300-600 nucleotides long DNA fragments in a single reaction. The main obstacle to sequence DNA fragments above this size limit is due to insufficient power of separation for resolving large DNA fragments that differ in length by only one nucleotide.

Microfluidic Sanger sequencing

Microfluidic Sanger sequencing is a chip based application of DNA sequencing. The sequencing steps such as thermal cycling, sample purification and capillary electrophoresis are all integrated on a chip using only nanoliter-scale sample volumes. This technology generates long and accurate sequence reads, while obviating many of the significant shortcomings of the conventional sequencing method by integrating and automating the standard sequencing steps.

Basically, in all high-throughput genome sequencing, fragmenting of the genome into small single-stranded pieces, followed by amplification of the fragments by Polymerase Chain Reaction (PCR) is the standard procedure. Adopting this procedure, each DNA fragment is irreversibly terminated with the incorporation of a fluorescently labeled dideoxy chain-terminating nucleotide, thereby producing a DNA ladder of fragments that each differ in length by one base and labelled with base-specific fluorescent at the terminal base. Amplified base ladders are then separated by Capillary Array Electrophoresis (CAE) with automated, *in situ* "finish-line" detection of the fluorescently labeled ssDNA fragments. The electrophoresis result will provide an ordered sequence of the fragments. These sequence reads are then computer assembled into overlapping or contiguous sequences which resemble the full genomic sequence once fully assembled.

Applications of microfluidic sequencing technologies

Other useful applications of DNA sequencing include

- single nucleotide polymorphism (SNP) detection,
- single-strand conformation polymorphism SSCP) heteroduplex analysis,
- short tandem repeat (STR) analysis.

Sequencing chemistry

- Thermal cycling (TC)

In the TC reaction chamber, dye-terminator sequencing reagent, template DNA and primers are loaded into the TC chamber and thermal-cycled for 35 cycles (at 95 °C for 12 seconds and at 60 °C for 55 seconds) will be carried out.

- Purification

The charged reaction mixture contains extension fragments, template DNA and excess sequencing reagent. It is conducted through a purification chamber at 30°C via a 33-Volts/cm electric field applied between capture outlet and inlet ports. The capture gel through which the sample is driven, consists of 40 μM of oligonucleotide complement to the primers will covalently bound to a polyacrylamide matrix. Extension fragments are immobilized by the gel matrix while excess primer, template, free nucleotides and salts are eluted through the capture waste port. The capture gel is heated to 67-75 °C to release extension fragments.

- Capillary electrophoresis

Extension fragments are injected into the CAE chamber where they are electrophoresed through a 125-167-V/cm field.

Platforms

The Apollo 100 platform (Microchip Biotechnologies Inc., Dublin, CA) integrates the first two Sanger sequencing steps which is thermal cycling and purification in a fully automated system. The Apollo 100 platform requires sub-microliter volumes of reagents.

Mutagenesis

Mutagenesis in the laboratory is an important technique whereby DNA mutations are deliberately engineered to produce mutant genes, proteins, strains of bacteria, or other genetically modified organisms. Various constituents of a gene, such as its control elements and its gene product, can be mutated so that the functioning of a gene or protein can be evaluated in detail. The mutation may also produce mutant proteins with different properties either enhanced or reduced or novel functions that may be of commercial use. Mutant strains may also be produced to have a practical application or allow the molecular basis of particular cell function. Currently development of the CRISPR/Cas9 technology are widely used.

Random mutagenesis

Initial approaches to mutagenesis rely on random mutations. Cells or organisms may be exposed to mutagens through UV radiation or mutagenic chemicals followed by selection of desired mutants. These methods may involve the use of doped nucleotides in oligonucleotides synthesis, or conducting a PCR reaction in conditions that enhance mismatched of nucleotides (error-prone PCR). The PCR products which contain mutation(s) are then cloned into an expression vector and the mutant proteins produced can then be characterized.

Site-directed mutagenesis

Many researchers seek to introduce some kind of changes to DNA in a site-specific manner. For that, analogs of nucleotides or chemicals such as aminopurine induces AT to GC transition, nitrosoguanidine, bisulfite and N^4-hydroxycytidine induce GC to AT transition. These techniques allow specific mutations to be engineered into a protein. Current techniques for site-specific mutation involve in using mutagenic oligonucleotides in a primer extension reaction with DNA polymerase. This methods allows for point mutation, deletion or insertion of small stretches of DNA to be introduced at targeted sites. While, the site-directed approach may be done systematically using such techniques, for example, alanine scanning mutagenesis is involved in alanine mutation, which is an important function of a protein.

Combinatorial mutagenesis

Combinatorial mutagenesis is a technique whereby large number of mutants screened for a particular characteristic. In this technique, a few selected positions or a short stretch of DNA will be modified to obtain a comprehensive library of mutant proteins. This technique involve with cutting a portion of DNA and replaced with a library of sequences containing all possible combinations at the desired mutation sites. The segment may involve at an enzyme active site or sequences that have structural significance or immunogenic property. A segment is allow to be inserted randomly into the gene in order to assess the structural or functional significance of a particular part of protein.

Insertional mutagenesis

Insertional mutagenesis using transposons, retrovirus such as mouse mammary tumor virus and murine leukemia virus can be used to identify genes involved in carcinogenesis and to understand the biological pathways of specific cancer. Various insertional mutagenesis techniques were successfully used to study the function of particular gene.

Homologous recombination

Homologous recombination can be used to produce specific mutation in an organism. Vector containing DNA sequence similar to the gene to be modified is introduced to the cell. Through recombination the target gene is replaced in the chromosome.

Gene synthesis

As the cost of DNA oligonucleotides synthesis is low, artificial synthesis of a complete gene is currently a do-able method for introducing mutations into gene. This method allows for extensive mutation on multiple sites, including the complete redesign of the codon usage of gene for a particular organism.

Genotyping

Genotyping is the process of determining the differences in the genetic make-up (genotype) of an individual by examining the individual's DNA sequence using biological assays and comparing it to another individual's sequence or a reference sequence. It reveals the alleles an individual has inherited from their parents.Traditionally genotyping is the use of DNA sequences to define biological populations by use of molecular tools. It does not usually involve defining the genes of an individual.Current methods of genotyping include restriction fragment length polymorphism identification (RFLP) of genomic DNA, random amplified polymorphic detection (RAPD) of genomic DNA, amplified fragment length polymorphism detection (AFLPD), polymerase chain reaction (PCR), DNA sequencing, allele specific oligonucleotide (ASO) probes and hybridization to DNA microarrays or beads. Genotyping is important in research of genes and gene variants associated with disease. Due to current technological limitations, almost all genotyping is partial. That is, only a small fraction of an individual's genotype is determined, such as with GBS (Genotyping by sequencing) or RADseq. Newmass-sequencing technologies promise to provide whole-genome genotyping (or whole genome sequencing) in the future.

Genotyping applies to a broad range of individuals, including microorganisms. For example, viruses and bacteria can be genotyped. Genotyping in this context may help in controlling the spreading of pathogens, by tracing the origin of outbreaks. This area is often referred to as molecular epidemiology or forensic microbiology.

Humans can also be genotyped. For example, when testing fatherhood or motherhood, scientists typically only need to examine 10 or 20 genomic regions (like single-nucleotide polymorphism (SNPs)). That is a tiny fraction of the human genome, which consists of three billion or so nucleotides.

When genotyping transgenic organisms, a single genomic region may be all that needs to be examined to determine the genotype. A single PCR assay is typically enough to genotype a transgenic mouse, the mouse is the mammalian model of choice for much of medical research today.

Qtl Mapping And Marker Assisted Selection

Marker assisted selection or marker aided selection (MAS) is an indirect selection process where a trait of interest is selected based on a marker in terms of morphological, biochemical or DNA/RNA variation which linked to a trait of interest (e.g. productivity, disease resistance, abiotic stress tolerance and quality), rather than on the trait itself. This process is used in plant and animal breeding.

For example, using MAS to select individuals with disease resistance involves identifying a marker allele that is linked with disease resistance rather than the level of disease resistance.

MAS can be useful to select for traits that are:

a) difficult or expensive to measure

b) exhibit low heritability and/or are expressed late in development

At certain points in the breeding process the specimens are examined to ensure that they express the desired trait.

Marker Types

Currently, the majority of MAS are DNA-based markers. Markers may be:

- **Morphological** - These markers are often detectable by eye and by simple visual inspection.
- **Biochemical**- A protein that can be extracted and observed.
- **Cytological** - The chromosomal banding produced by different stains.

Positive And Negative Selectable Marker

The following terms are generally less relevant to discuss about MAS in plant and animal breeding, but are highly relevant in molecular biology research:

- **Positive** selectable markers are selectable markers that confer selective advantage to the host organism.
- **Negative** selectable markers are selectable markers that eliminate or inhibit growth of the host organism upon selection.

A distinction can be made between selectable markers and screenable markers as most MAS uses screenable markers rather than selectable markers.

Gene Versus Marker

GENE	MARKER
Involve in the production of proteins or RNA	A DNA sequence, A morphological markers, A biochemical markers
Produce a desired trait or phenotype	Will move together during segregation of gametes due to their proximity on the same chromosome and concomitant reduction in re-combination

Will move together during segregation of gametes due to their proximity on the same chromosome and concomitant reduction in recombination	If the markers is not known, gene of the interest cannot be identified and some inaccurate results due to inaccurate tests will occurred
If the gene of interest is not known, markers linked to the gene of interest can still be used to select for individuals with desirable alleles of the gene of interest	False positive; Only perfect marker will give a significant result

Characteristics of an ideal marker

- Easy recognition of all possible phenotypes from all different alleles
- Demonstrates early measurable differences in expression between trait types or gene of interest alleles
- Should not have variable success depending on the allele at the marker locus or the allele at the target locus
- Low or null interaction among the markers allowing the use of many at the same time in a segregating population
- Abundant in number
- Polymorphic

Morphological markers are associated with several general deficits that reduce their usefulness. Those that was identified was including:

- the delay of marker expression
- dominance
- deleterious effects
- pleiotropy
- confounding effects of genes unrelated to the gene or trait of interest but which also affect the morphological marker (epistasis)
- rare polymorphism
- frequent confounding effects of environmental factors which affect the morphological characteristics of the organism

To avoid problems specific to morphological markers, the DNA-based markers have been developed.

Advantages of DNA markers

- they are highly polymorphic,
- exhibit simple inheritance (often codominant),
- abundant throughout the genome,
- easy and fast to detect,
- exhibit minimum pleiotropic effects,
- detection is not dependent on the developmental stage of the organism.

Numerous markers have been mapped to different chromosomes in several organisms including

- rice,
- wheat,
- maize,
- soybean,
- livestock such as cattle, pigs and chickens.

Those markers have been used in diversity analysis, parentage detection, DNA fingerprinting, prediction of hybrid performance, in indirect selection processes and manual selection of individuals for further propagation.

Major genes selection that linked to markers are based on

- disease resistance,
- male sterility,
- self-incompatibility and others related to
- shape, color and architecture of whole organisms.

There are several indications for the use of molecular markers in the selection of a genetic trait. In situation such as:

- the selected character is expressed late in plant development
- the expression of the target gene is recessive
- there is requirement for the presence of special conditions in order to invoke expression of the target gene(s)

- the phenotype is affected by two or more unlinked genes (epistasis)
- the cost of genotyping is reducing while the cost of phenotyping is increasing.

Steps involve in MAS

- map the gene or quantitative trait locus (QTL) of interest by using different techniques
- then using this information for marker assisted selection

Generally, the markers to be used should be close to gene of interest (<5 recombination unit or cM) in order to ensure that only minor fraction of the selected individuals will be recombinants. Usually, not only a single marker but rather two markers are used in order to reduce the chances of an error due to homologous recombination.

Techniques

In plants, QTL mapping is generally achieved using bi-parental cross populations. It is cross between two parents which have a totally different phenotype for the trait of interest that are developed. Commonly used populations are

- recombinant inbred lines (RILs),
- doubled haploids (DH),
- back cross and F_2.

Linkage between the phenotype and markers which have already been mapped is tested in these populations in order to determine the position of the targeted QTL. Such techniques are based on linkage and are therefore referred to as "linkage mapping".

Single Step MAS and QTL Mapping

A single-step method for breeding typical plant populations has been developed. In such an approach, in the first few breeding cycles, markers linked to the trait of interest are identified by QTL mapping and later the same information is used in the same population. In this approach, pedigree structure is created from families that are created by crossing number of parents in three-way or four way crosses. Both phenotyping and genotyping is done using molecular markers that mapped the possible location of QTL of interest. This will identify markers and their favourable alleles. Once these positive marker alleles are identified, the frequency of such alleles will be increased and response to marker assisted selection estimation. Marker allele(s) with desirable effect will be further used in next selection cycle or in subsequent experiment.

High-Throughput Genotyping Techniques

Recently high-throughput genotyping techniques are developed to allows marker aided screening on many genotypes. This will help breeders in shifting traditional breeding to marker aided selection. One example of such automation is using DNA isolation robots, capillary electrophoresis and pipetting robots.

Mas For Backcross Breedinga

Minimum of five or six-backcross generations are required to transfer a gene of interest from a donor to a recipient. The recovery of the recurrent genotype can be accelerated with the use of molecular markers. If the F1 is heterozygous for the marker locus, individuals with the recurrent parent allele(s) at the marker locus in first or subsequent backcross generations will also carry a chromosome tagged by the marker.

Marker Assisted Gene Pyramiding

Gene pyramiding can be applied to enhance resistance to disease and insects by selecting for two or more than two genes at a time. For example, in rice such pyramids have been developed against bacterial blight and blast. The advantage of the usage of the markers in this case allows to select for QTL-allele-linked markers that have same phenotypic effect. MAS has also been proved useful for livestock improvement.

Quantitative Traits

A quantitative trait locus (QTL) is a section of DNA that correlates with variation in a phenotype. Usually the QTL is linked to the genes that control the phenotype. QTLs are mapped by identifying molecular markers such as SNPs or AFLPs which correlate with an observed trait. This is often an early step in identifying and sequencing the actual genes that cause the trait variation. Quantitative traits are phenotypes characteristics that vary in degree and can be attributed to polygenic effects and their environment. Polygenic inheritance refers to inheritance of a phenotypic characteristic that is involve with to two or more genes and are quantitatively measurable. Multifactorial inheritance refers to polygenic inheritance that also interact with the environment. Unlike monogenic traits, polygenic traits do not follow patterns of Mendelian inheritance. Instead, their phenotypes typically vary along a continuous gradient depicted by a bell curve.

Multifactorial traits

Traits are usually controlled by genetic factors and the environment. Usually, multifactorial traits in human organisms are such as height, skin color and body mass. All of these phenotypes are complicated by a great deal of genes and environmental effects. The continuous distribution of traits such as height and skin color reflects the action of genes that do not manifest typical patterns of dominance and recessiveness. Instead the contributions of each involved locus are thought to be additive. Researchers have distinguished this kind of inheritance as polygenic, or quantitative inheritance. Thus, due to the nature of polygenic traits, inheritance will not follow the same pattern as a simple monohybrid or dihybrid cross. Polygenic inheritance can be explained by Mendelian inheritance at many loci, resulting in a trait which is distributed normally. If n is the number of involved in loci, then the coefficients of the binomial expansion of $(a + b)^{2n}$ will give the frequency of distribution of all n allele combinations. For a sufficiently high n, this binomial distribution will begin to resemble a normal distribution. This shows that, a disease state will become apparent at one of the tails of the distribution, which past some threshold value. Disease states of increasing severity will be expected to be the further one goes past the threshold and away from the mean.

Heritable disease and multifactorial inheritance

A mutation resulting in a disease state is often recessive, so both alleles must be mutant in order for the disease to be expressed phenotypically. A disease or syndrome may also be the result of the expression of mutant alleles at more than one locus. When more than one gene is involved, with or without the presence of environmental triggers, it is considered that the disease is the result of multifactorial inheritance.

The more genes involved in the cross, the more the distribution of the genotypes will resemble a normal, or Gaussian distribution. This shows that multifactorial inheritance is polygenic and genetic frequencies can be predicted by polyhybrid Mendelian cross. The paradigm of polygenic inheritance as being used to define multifactorial disease has encountered much disagreement.

The assumption of polygenic inheritance is that all involved loci make an equal contribution to the symptoms of the disease. This should result in a normal curve distribution of genotypes. If it is not, the idea of polygenetic inheritance cannot be supported for that illness.

QTL Mapping

A quantitative trait locus (QTL) is a region of DNA that is associated with a particular phenotypic trait. These QTLs are often found on different chromosomes. Knowing the number of QTLs that explains variation in the phenotypic trait tells us about the genetic architecture of a trait. It may tell us that plant height is controlled by many genes of small effect, or by a few genes of large effect. Another use of QTLs is to identify candidate genes underlying a trait. Once a region of DNA is identified as contributing to a phenotype, it can be sequenced. The DNA sequence of any genes in this region can then be compared to a database of DNA for genes whose function is already known.

In a recent development, classical QTL analyses are combined with gene expression profiling such as by DNA microarrays. Such expression QTLs (eQTLs) describe cis- and trans-controlling elements for the expression of disease-associated genes. Observed epistatic effects have been found beneficial to identify the gene responsible by a cross-validation of genes within the interacting loci with metabolic pathway and scientific literature databases.

For organisms whose genomes are known, statistical geneticists try to exclude genes in the identified region whose function is known with some certainty not to be connected with the trait in question. If the genome is not available, it may be an option to sequence the identified region and determine the putative functions of genes by their similarity to genes with known function from other genomes. This can be done using BLAST analysis which is available online. This tool allows users to enter a primary sequence and search for similar sequences within the BLAST database of genes from various organisms. It is often not the actual gene underlying the phenotypic trait, but rather a region of DNA that is closely linked with the gene. Besides that, statistical geneticists uses QTL mapping to determine the complexity of the genetic architecture underlying a phenotypic trait. For example, they may be interested in knowing whether a phenotype is shaped by many independent loci, or by a few loci and how do those loci interact. This will provide information on how the phenotype are evolving.

Analysis of variance

The simplest statistic method for QTL mapping is analysis of variance ANOVA, also known as marker regression at the marker loci. In this analysis, during a backcross user can analyse a t-statistic to compare the averages of the two marker genotype groups. For other types of crosses such as the intercross where there are more than two possible genotypes, user can use more general form of ANOVA, which is the F-statistic. The ANOVA approach for QTL mapping has three important weaknesses

- Lack of separate estimation on QTL location and QTL effect. QTL location is indicated only by looking at which markers give the greatest differences between genotype group averages and the apparent QTL effect at a marker will be smaller than the true QTL effect as a result of recombination between the marker and the QTL.
- It is a must to discard individuals whose genotypes are missing at the marker.
- For markers that are widely spaced, the QTL will be quite far from all markers and the power for QTL detection will decrease.

Interval mapping

The term 'interval mapping' is used for estimating the position of a QTL within two markers. Interval mapping is originally based on the maximum likelihood but there are also very good with simple regression. Interval mapping overcomes the three disadvantages of analysis of variance at marker loci. The method makes use of a genetic map of the typed markers and assumes the presence of a single QTL. In interval mapping, each locus is considered as one at a time and the logarithm of the odds ratio (LOD score) is calculated for the true QTL. The odds ratio is related to the Pearson correlation coefficient between the phenotype and the marker genotype for each individual in the cross experiment.

The principle for QTL mapping is:

1) The Likelihood can be calculated for a given set of parameters particularly QTL effect and QTL position given the observed data on phenotypes and marker genotypes
2) The estimates for the parameters are those where the likelihood are the highest
3) A significant threshold can be established by permutation testing

Conventional methods for the detection of quantitative trait loci (QTLs) are based on a comparison of a single QTL models with a model without QTL. For instance, in the interval mapping method, the likelihood for a single putative QTL is assessed at each location on the genome. However, QTLs located elsewhere on the genome can have an interfering effect too. As a consequences, the power of detection may be compromised and the estimation of locations and effects of QTLs will be biased. In addition, non-existing so-called ghost QTLs may appear too. Therefore, it is obvious that multiple QTLs could be mapped more efficiently and more accurately by using multiple QTL models. A popular approach to handle QTL mapping is by scanning the genome with multiple QTL that contribute to a trait and add a known QTL to the regression model as QTLs. This method, termed as composite interval mapping as both the location and effects size of QTL are more accurately determined than single-QTL approaches.

Composite interval mapping (CIM)

In CIM, interval mapping will be performed using a subset of marker loci as covariates. These markers serve as proxies for other QTLs, increases the resolution of interval mapping, linked QTLs and reducing the residual variation. The key problem with CIM is the choice of suitable marker loci to serve as covariates because once these have been chosen, CIM turns the model selection problem into a single-dimensional scan. The appropriate markers are those closest to the true QTLs and so if one could not find these, the QTL mapping problem would be complete anyway.

Family-pedigree based mapping

Family based QTL mapping is the only way for mapping of genes where crosses are difficult to make. Here, linkage and association mapping involves multiple families instead of a single family.

Animal Genotyping

The Ion Proton™ and Ion PGM™ Sequencers enable plant and animal researchers to use flexible, high-throughput GBS to perform fast and cost-effective studies. This tool is useful for association studies and genomics-assisted breeding.

GBS: advantages of the Ion Proton™ and Ion PGM™ Sequencers

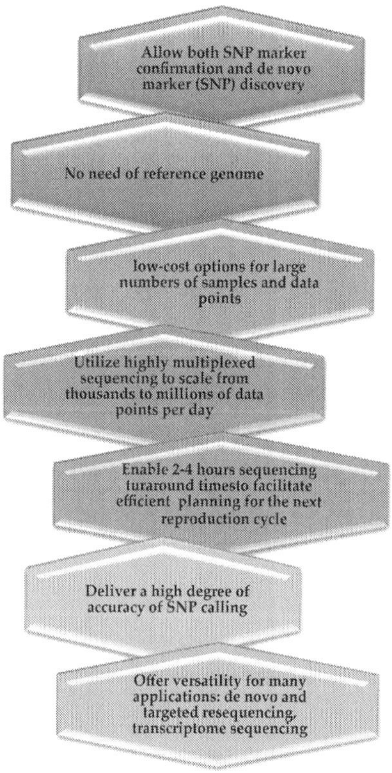

Multiplex enrichment PCR

In cases where a set of SNPs has been defined for a section of the genome, this approach uses PCR primers designed to amplify the targets of interest. Creating and ordering custom panels is easy with the Ion AmpliSeq™ Designer: a free, online assay design tool that enables custom design for any genome. Use the preloaded bovine, chicken, pig and sheep genomes to design your custom panel, or easily upload a FASTA file with your desired reference sequences for your genome of interest.

Genotyping by Sequencing Workflow

- Library construction using Ion ampliseq Panels on the Ion Chef System
- Template Preparation using Ion Chef System
- Sequence Run Ion S5 or Ion S5 XL Systems
- Analyze Data Torrent Suite Software

CHAPTER

6 TOOLS AND TECHNIQUES IN INTERPRETING BIOLOGICAL DATA

COMPUTATIONAL TOOLS

Some examples of software packages that can be used in analysing sequencing data:

- Agent-Based Simulations (NetLogo, Swarm)
- Bayesian Analyses (OpenBugs, MCSim, *DoseSim*)
- Chemical Information System Software (Mesa Analytics)
- Computer-Aided Drug Design (OpenEye Suite)
- General Programming and Numerical Analysis (Python, NumPy, SciPy, Mathematica, Maple, Octave, MATLAB, Fortran, C, C++, MPI Libraries)
- Gene Sequence Analysis (SHRiMP, ABI SOLiD)
- Microarray Analysis (ArrayTrack)
- Molecular Biology (EMBOSS)
- Molecular Docking (AutoDock)
- Molecular Dynamics (UCSF Chimera, GROMACS, NAMD,

Amber, Tinker, LAMMPS)

- Pathway Analysis (Ingenuity Pathway Analysis, GeneGo MetaCore)
- Pharmacokinetic Analysis (MCSim, Berkeley Madonna, acslXtreme, WinNonLin)
- Phylogenetic Analysis (Phylip, MrBayes)
- Population & Migration Analysis (Migrate)
- Quantum Chemistry / Electronic Structure Theory (Gaussian, AMPAC, GAMESS)
- Reaction Diffusion Simulations (Smoldyn, MesoRD)
- Sequence Alignment (clustaw, HMMR, Wise2, mpiblast, blastall)
- Statistical Analysis (R, SAS, Bioconductor)

DNA TOOLS

Below are some of the tools that can used for DNA research

- BBOCUS by Ferro and Purrello lab
    - BBOCUS (Back Translation Based On Codon Usage Strategy) is used to backtranslate an amino acid in a protein.
- Benchling by Benchling, Inc.
    - It is a free online tools for vector editing, restriction analysis, primer search, multi-sequence alignment, and more.
    - Completely browser-based and no installation are required.
    - Allows for easy collaboration on sequences between researchers via shared libraries and/or links.
    - Can also generate vector plasmid maps for presentations and publications.
    - Imports from a wide variety of formats including genbank, fasta, ApE, DNA Strider, lasergene, and more.

    - Include tracks complete version, history of editing and cloning steps.

- Biopolymer calculator by Schepartz lab
    - Calculate extinction coefficients, Tm's, and base composition for your DNA or RNA; calculate amino acid composition and extinction coefficient for your protein.
- Clipboard by Austin Che
    - Web tool for getting complement, reverse complement, translation and restriction enzyme analysis of a DNA sequence.
- Cytostudio by Molecula Maxima
    - Is an integrated development environment and a compiler for a high-level bio-programming language for Synthetic Biology.
    - Contain PDB IDs, GenBank accession numbers, FASTA format files and codon usage optimizations are part of the language.
    - Visualize plasmid variables (circular view) and strand variables (linear view) as composites of iGEM parts and devices.
    - Proteins Viewer.
    - Use embedded web views to easily search in common databases.
    - Write plugins with JavaScript or Visual Basic to fit Cytostudio to your needs.
- DNAWorks by Hoover and Lubkowski
    - A web tool for optimizing melting temperature during gene synthesis.
- File format converter by Austin Che

    - Web tool for converting between sequence file formats.
- Geneious by Biomatters
    - Comprehensive suite of tools for molecular biology.
    - Drag and drop import a vast range of file formats.
    - Simulate Gibson Assembly, TOPO Cloning and other cloning operations in one easy step.
    - Primer design and testing powered by Primer3.
    - Built in codon optimization.
    - Immediate access to Golden Gate, Biobrick and other important restriction enzymes.
    - Automatic plasmid annotation.
    - Automatic tracking of cloning operations.
- Genome Compiler by Genome Compiler
    - Is the industry's most user friendly genetic engineering design tool. It allows users to manipulate genetic information from genes to plasmids to whole genomes.
    - Users can rapidly access extensive libraries of genetic parts, and easily order the design from a variety of providers.
    - Some of the key features include: a user-friendly workspace for handling genetic constructs of all sizes.
    - Consists a powerful gene search capability with NCBI access,
    - Contain a built-in constructs library that includes a parts registry, vectors-plasmids, complete genomes, as well as seamless collaboration through the Genome Compiler Cloud.
    - Users can drag components between layers and projects (in multiple tabs), design genetic constructs across tight-

ly integrated layers, operon/part level, amino acid, and DNA, perform re-annotation and reverse complement, employ automatic sequence detection to point out restriction enzymes and protein sub-domains.

- GeneDesign by Boeke lab
    - Collection of online (and some command line) tools for codon optimization and shuffling, restriction site editing, and so on.
- GeneDesigner by DNA2.0
    - Combine genetic building blocks by drag-and-drop, codon optimize, restriction site editing, sequence oligo design.
    - Free to download and works on Mac or PC.
    - User agreement is somewhat restrictive (i.e., does not allow you to sell genes designed on the program without permission).
- GenoCAD
    - GenoCAD is a design tool that uses collections or libraries of genetic parts.
    - Allow to design a DNA sequence and allow to be exported using standard formats.
- NEB Cutter by New England Biolabs, Inc.
    - Tool for finding restriction sites.
- Synthetic Gene Designer by Gang Wu
    - A web platform that allows codon optimization to various extent.
    - Compatible with non-standard genetic codes.
    - Integrate CUTG codon.
    - Restriction site avoidance.

    - Allow user to design oligo.
- Vector NTI by Informax, Inc.
    - Free-to-academics tool for sequence analysis and data management.
- VectorEditor online tools
    - VectorEditor: a visual DNA editing and annotation tool.
- Molecular Cloning Designer Simulator by Zhenyu Shi (Latest update 2016-05-13) Published at MEC
    - MCDS aims to set up a standard way for designing, describing and simulating molecular cloning and genetic engineering.
    - MCDS can design primers, simulate restriction enzyme digestion, gel electrophoresis, Gateway cloning, BioBrick assembly, Gibson assembly, yeast assembly, site-specific recombination and assembly, lambda red knockout/knocking, homologous recombination, CRISPR digestion.
    - MCDS allows users to customize restriction enzymes, site-specific recombination sites, gRNAs.
    - MCDS will able to manage and track every molecular cloning and genetic engineering operation in a unique flowchart, which can be printed or exported as emf vector image. The flowchart of MCDS includes all necessary information including PCR primers, product length restriction enzyme sites and even digestion buffers (users can customize support for digestion buffers).
    - MCDS supports GeneBank formats, AB1 formats (and can display sequencing data) and provides embedded access to NCBI and KEGG.

RNA TOOLS

- Appendix by Ambion, Inc.

    - It is a website designed with many useful nucleic acid parameters.
- mFold by Michael Zuker.
    - It is a page with links to sites for predicting RNA and DNA folds, calculating Tm's and free energies.

PROTEIN TOOLS

- Cn3D by NCBI
    - Allows users to view 3D structures from NCBI's Entrez retrieval service.
    - It doesn't read PDB files
    - User friendly than DeepView.
- DeepView by GlaxoSmithKline & Swiss Institute of Bioinformatics
    - Can be used for viewing and studying protein structure.
- ExPASy Proteomics server by the Swiss Institute of Bioinformatics
    - Collection of links to many pages to calculate parameters of your favorite proteins.
- Modeller by Sali Lab
    - Can be used for homology or comparative modeling of protein three-dimensional structures.
- Zinc Finger Tools by Barbas Lab
    - User can design Zinc Finger DNA binding proteins.

CAD TOOLS

- TinkerCell by Deepak Chandran
    - Was designed to bridge computational and experimental

biology.

    - Users can construct computational models using biological parts, cells, and modules.
    - Users can Plug-in interface allows contributors to add new features and functions (C and Python).
- Metabolic Tinker by Kent McClymont and Orkun Soyer
    - Construct thermodynamically feasible metabolic paths among user-defined compounds.
    - Uses all available metabolic data (updated regularly).
    - Source code available for further development.

GENERAL TOOLS

- Colibri by Institut Pasteur
    - It is a *E. coli* genome site.
    - Users can get sequences.
    - Users can locate the position of the gene on interest in the chromosome.
    - Identified the function of your gene.
    - You can also search for protein sequences/motifs within the *E. coli* genome.
- Registry of Standard Biological Parts by MIT
    - Open repository of Bio Bricks- the place for all your standard biological parts.
        - PaR-PaR Laboratory Automation Platform.
            - PaR-PaR allows users to use liquid-handling robots effectively.

CHAPTER

7 CASE STUDIES

CASE STUDY 1

TOOLS AND TECHNIQUES -SAMPLE EXPERIMENT

ISOLATION AND DESCRIPTION OF PARTIAL HOMOGENTISATE PHYTYLTRANSFERASE GENE FROM ELAEIS GUINEENSIS (MALAYSIA)

INTRODUCTION

α-tocopherol (Vitamin E) are an essential nutrient and universal constituents in all higher plants. It consists of a polar chromanol group and a non-polar isoprenoid side chain that is fully saturated. The committed step in the α-tocopherol biosynthesis basically involves the condensation of homogentisic acid (HGA) and phytyldiphosphate (PDP) catalysed by the membrane bound homogentisate phytyltransferase (HPT). The substrate, HGA, is produced by deamination and oxygenation of tyrosine into p-hydroxyphenylpyruvate (HPP). Homogentisate phytyltransferase catalyzed reaction produces MPBQ (2-methyl-6-phytyl-1,4-benzoquinol) which is the first intermediate product for the production of vitamin E. It will be further cyclised and methylated by subsequent enzymes to produce α-tocopherol and α-tocotrienol isoforms.

Homogentisate phytyltransferase is a chloroplast enzyme and is present in a relatively high concentrations in leaf tissues as vitamin E synthesis occurs exclusively in the plastids of higher plants. Homogentisate phytyltransferase has been widely isolated from *Glycine max* (NM001250971), *Solanum pennellii* (HQ 014373.1), *Solanum tuberosum* (HM217350.1), *Arabidopsis thaliana* (NM001161137.1), *Malus domestica* (GQ329852.1), *Lactuca sativa* (FJ194492.1), *Cuphea pulcherrima* (DQ231058.1), *Allium porrum* (DQ231057.1), *Coriandrum sativum* (EU617358.1), *Manihot esculenta* (EU606022.1) *Zea mays* (NM001112407.1) and *Triticum aestivum* (DQ231056.1).

Highly identical amino acids residues at analogous parts of proteins are often required for basic cellular function and stability. With this in mind, rather than isolating the full length cDNA, the conserved region of HPT that is similar in local nucleotide sequence/structure shared by multiple plant species was isolated from oil palm (*Elaeis guineensis*) known to have high content of vitamin E. The partial sequence obtained was subjected to bioinformatics analysis and the region was found to share a high level of sequence homology and have identical regional structural characteristics as the protein in the full length.

Studies in isolating the conserved region for characterization purposes have been done. This study was attempted to isolate part of the conserved region of HPT gene and to carry out bioinformatics analysis to obtain an insight on its structural and functional properties and will thus confirm that it is a paralog to the sequences from other plant species found in GenBank.

MATERIALS AND METHODS

Plant material

Six-week-old young seedlings of Elaeis guineensis leaf were obtained from Golden Hope, Banting, Selangor (Figure 1).

Figure 1: Six week- old young seedling of *Elaeis guineensis* leaves.

Plasmid, bacterial strains and growth

Plasmid yT&A (EPICENTRE Biotechnologies, USA) was used for standard DNA manipulation. It consists of T7 promoter with LacZ, ampicillin resistant (Amp^r) and multiple cloning sites (MCS) consisting of KpnI, SmaI, EcoRI, BglII, BamHI, XbaI, SalI, PstI and HindIII (Figure 2) . DNA fragment was ligated in between EcoRI and BglII. *Escherichia coli* XL1-Blue ECOS™101(EPICENTRE Biotechnologies, USA) (endA1 recA1 relA1 gyrA96 hsdR17 (r_k^-,m_k^+) phoA supE44 thi-1 Δ(lacZYA-argF)U169 Φ80Δ(lacZ)M15F$^-$) was used for transformation purposes. Plate and liquid culture of *Escherichia coli* DH5α was grown from a single colony at 37°C for 16 h. For liquid culture, it was grown with vigorous shaking at 220 rpm. All types of growth media for Escherichia coli contains ampicillin (50 μg/mL).

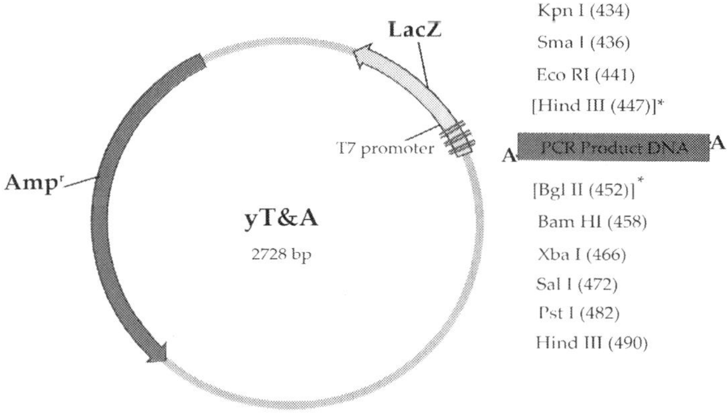

Figure 2: Schematic diagram of the yT&A. The HPT fragment was inserted in between HindIII and BglII. Multiple cloning sites indicated as MCS.

Isolation of RNA

Leaf samples were transferred into prechilled mortars and pestles and were ground to a very fine powder with liquid nitrogen. The finely powdered sample was transferred to a fresh 1.5 mL microfuge tubes using chilled spatula. Eight hundred microlitre of prewarmed (65°C) extraction buffer was added and vortexed immediately and was incubated for 10 min at room temperature with a intermittent mixing at ever 2 min for about 4 sec. The solution was centrifuged at 12 000 rpm for 10 min at 4°C and the supernatant was transferred to new tubes. One per nine part of sodium acetate was added and mixed by flicking. One volume of cold isopropanol was added and was mixed thoroughly by inverting. The solution was centrifuged again at 12 000 rpm for 10 min at 4°C. The supernatant was discarded and the pellets were completely re-suspended in 750 μL ddH_2O and the tube was placed on ice for 1 min. Two hundred fifty microlitres of 9 M LiCl (2.25 M final concentration) was added and the solution was mixed thoroughly by inverting. The tubes was placed on ice and incubated at 4°C overnight. The precipitate RNA was collected by centrifugation at 14 000 rpm for 20 min at 4°C. The supernatant was removed by pipetting. RNA pellet was dissolved by pipetting in 30 μL dH_2O and was incubated overnight at 4°C.

RNA analysis

The yield and quality of total RNA were measured by absorbance at 230, 260 and 280 nm ($A_{260/230}/A_{260/280}$) using Nano Drop spectrophotometer (Nano Drop, Techologies Inc.) and was separated using 1% agarose gel electrophoresis.

Agarose gel electrophoresis

Agarose gel was prepared. The electrophoresis buffer 1X TAE was prepared from a 10X TAE. Aliquots of 4 μL of any types of samples (plasmid or DNA or digestion sample) and 1 μL of loading dye were mixed and loaded into the wells of the gel. Electrophoresis was carried out at constant voltage of 80V for 45 min. Deoxynucleotide acid staining was carried out for 15 min using ethidium bromide staining solutionand destaining was carried out for 30 min using sterilized distilled water. The gel photograph of the gel was taken using the gel documentation system under the emission of U.V light.

Isolation of short cDNA of HPT gene

The isolated RNA was used to perform first strand cDNA synthesis by using Exact START Eukaryotic mRNA 5`- & 3`-RACE kit from EPICENTRE Biotechnologies, WI, USA.

Alkaline phosphatase treatment

The following reaction was assembled 80 μL nuclease free water, 10 μL Apex Reaction buffer, 5 μL total RNA sample (1-10 μg) 5 μL Apex Heat-Labile Alkaline Phosphatase. The reaction volume was top up to 100 μL. The reaction was incubated at 37°C for 15 min. Phenol: chloroform extraction was carried out by adding 1:1 of phenol and chloroform in the 100 μL reaction. The reaction was vortexed vigorously and centrifuged at 12 000 rpm for 1 min to separate two phase. The upper phase was collected into a new sterile tube. The remaining lower phase was subjected for second phenol: chloroform extraction. Two hundred microlitre of chloroform: isoamylalcohol (24:1) was added into 200 μL of the extracted phenol: chloroform reaction containing RNA. The reaction was vortexed vigorously and was centrifuged for 1 min. The upper phase was collected into new sterile tube. Twenty microlitre of 3 M sodium acetate and 100 μL of isopropyl alcohol were added into the aqueous phase containing RNA. The reaction was vortexed vigorously prior incubation in ice for 15 min. Then the aqueous was centrifuged at 12 000 rpm for 15 min at 4°C. The supernatant was removed without disturbing the RNA (pellet). The pellet was washed twice with 100% cold absolute ethanol. The ethanol was discard by centrifugation at 12 000 rpm for 15 min at 4°C. The pellet was air dried before dissolving it in 7.5 μL nuclease free water.

Tobacco acid pyrophosphatase treatment

One microlitre of TAP buffer, 0.5 μL ScriptGuard RNase inhibitor, 7.5 μL RNA sample (1-10 μg), 1 μL TAP enzyme was assembled on ice in a thin walled PCR tube. The reaction was incubated at 37°C for 30 min.

5`RACE aceptor oligo ligation

Ten microlitre of TAP-treated RNA from earlier method, 4 μL nuclease free water, 2 μL RNA ligase buffer, 1 μL TAP STOP buffer, 5`-RACE Acceptor Oligo, 2 mM ATP solution, T4 RNA ligase was assembled on ice. Each reagent was pipetted several times thoroughly before it was added into the reaction tube. The reagents were added in order. TAP STOP buffer was added before adding 2 mM ATP solution. The reaction was incubated at 37°C for 30 min.

First-strand cDNA synthesis

Twenty microlitre of 5` RACE Acceptor Oligo-ligated RNA, 14 μL of nuclease free water, 1 μL cDNA synthesis primer, 2 μL dNTP PreMix, 2 μL MMLV RT Buffer, 1 μL MMLV reverse transcriptase was assembled. The reaction was incubated at 37°C for 1 h followed by 85°C for 10 min and at 55°C for 5 min after adding 1 μL of RNase solution. The reaction was kept until the following step.

Second-strand cDNAsynthesis

Forty one microlitres of nuclease treated first strand cDNA, 18 μL nuclease free water, 5 μL of PCR primer 1, 5 μL PCR primer 2, 30 μL FailSafe PCR Premix E and 1 μL thermostable DNA polymerase with proofreading (2.5 U) was assembled. Each reagent was pipetted several times thoroughly before it was added into the reaction tube at 55°C. The PCR cycle was carried out at 95°C for 30 sec for initial denaturation step and 21 cycles were run under following conditions: 95°C for 20 sec, 60°C for 20 sec, 72°C for 3 min and was hold at 4°C. The double stranded DNA was stored at -20°C.

Homogentisate phytyltransferase amplification

Upon synthesis of second strand cDNA synthesis HPT gene was amplified using primers designed to pick up HPT gene. Figure 3 shows the conserved amino acid sequences which were represented by identical letters was aligned. Conserved amino acid sequences are marked by strings of * below the third line of the sequence alignment. The primers for amplification were designed based on the output of the nucleotide blast analysis. These primers Zea (F1) and ZEA (R3) that flanked the conserved region based on the pairwise sequence alignment were synthesized by First Base Company, Singapore. The primer sequences are listed in the Table 1.

```
Sequence alignment

Consensus key (see documentation for details)
* - single, fully conserved residue
: - conservation of strong groups
. - conservation of weak groups
  - no consensus

Name : BWB12225      Len : 399  Zea mays
Name : BWB11818      Len : 400  Triticum aestivum
Name : BWB11285      Len : 395  Allium ampeloprasium

CLUSTAL W (1.81) multiple sequence alignment

BWB12225    ---MDALRLRPSLLSVRPG--AARPRDHFLPPCCSIQRNGEGRICFSSQRTQGPTLHHHQ
BWB11818    ---MDSLRLRPSSLRSAPGAAAARRRDHILPSFCSIQRNGKGRGRVTLSIQASKGPTINHCK
BWB11285    MLSMDSLLTKPVVIPLPS---PVSLPTILRGSSAPGQYSCRNYNPIRIQRCLVNYEHVKP
               **:*  :*  :     .    ..      :  . ..*  . ..   :  *      :

BWB12225    KFFEWKSSYCRISHRSLNTSVNASGQQLQSEPETHDSTTIWRAISSLDAFYRFSRPHTV
BWB11818    KFFEWKSSYCRISHRSLNTSVNASGQQLQSEPETHDSTTIWRAISSLDAFYRFSRPHTV
BWB11285    RFTTCSRS-QKLGHVKATS----EHSLESGSEGYTPRSIWEAVLASLNVLYKFSRPHTI
            :*    . *  ::.*  . .:       :.*:.  .* :  . ::*..: :**:.:*:******:

BWB12225    IGTALSIVSVSLLAVQSLSDISPLFLTGLLEAVVAALFMNIYIVGLNQLFDIEIDKVNKP
BWB11818    IGTALSIVSVSLLAVQSLSDISPLFLTGLLEAVVAALFMNIYIVGLNQLFDIEIDKVNKP
BWB11285    IGTALSIVSVSLLAVQSLSDISPLFLTGLLEAVVAALFMNIYIVGLNQLFDIEIDKVNKP
            ****:.*:*****.*:*********:.*************:*********************

BWB12225    TLPLASGEYTPATGVAIVSVFAAMSFGLGWAVGSQPLFWALFISFVLGTAYSINLPYLRW
BWB11818    TLPLASGEYTPATGVAIVSVFAAMSFGLGWAVGSQPLFWALFISFVLGTAYSINLPYLRW
BWB11285    DLPLASGEYSPRAGTAIVIASAIMSFGIGWLVGSWPLFWALFISFVLGTAYSINLPFLRW
             ********:* :*.***  .  * ****:** *** *******************:***:**

BWB12225    KRFAVVAALCILAVRAVIVQLAFFLHIQTFVFFRRPAVFSRPLLFATGNTFFSVVIALFK
BWB11818    KRSAVVAALCILAVRAVIVQLAFFLHIQTFVRRPAVFSKPLIFATAFMTFFSVVIALFK
BWB11285    KRSAVVAAICILAVRAVIVQLAFFLHIQSFVFKRPASFTRPLIFATAFMSFFSVVIALFK
            ** *****:*********************:***:***  *::**:***.**:**********

BWB12225    DIPDIEGDRIFGIRSFSVRLGQKKVFWICVGLLEMAYSVAILMGATSSCLWSKTATIAGH
BWB11818    DIPDIEGDRIFGIQSFSVRLGQSKVFWTCVGLLEVAYGVAILMGVTSSSLWSKSLTVVGH
BWB11285    DIPDIDGDKIFGIHSFSVRLGQERVFWICIYLLEMAYTVVMVVGATSSCKWSJCKVIGH
            *****:**:****:********* .:*** *: ***:** *.:::*.***.****   *: **

BWB12225    SILAAILWSCARSVDLTSKAAITSFYMFIWKLFYAEYLLIPLVR
BWB11818    AILASILWSSARSIDLTSKAAITSFYMLIWRLFYAEYLLIPLVR
BWB11285    AILGSLLWNRARSHGPMTKTTITSFYMFVWKLFYAEYLLIPFVR
            :**.::**. ***  .   :*::******::*:**********:**
```

Figure 3: The fasta format of a multiple sequence alignment of three HPT protein sequences. Sequence identification appears at the beginning of each row and the correspond names of the sequences are listed at the top. The level of sequence conservation is shown as being the least conserved (.) or (:) and being the most conserved as (*).

Table 1: PCR and DNA sequencing primers used in this study.

Primer name	T_m °C	Nucleotide sequence 5′ to 3′	template
Zea (F1)	60.5	ATGGACGCGCTTCGCCTAC	cDNA
ZEA (R3)	59.7	TCACCGCACCAGAGGGATG	cDNA

Polymerase chain reaction was carried out with 1 U/μL of Platinum® Taq DNA Polymerase High Fidelity (5 U/μL, Invitrogen™, NY, USA) in a 25 μL reaction mixture. 10 μM of each primer (Zea (F1) and ZEA (R3)), 10 mM dNTP mixture, 50 mM $MgSO_4$ and Platinum® Taq DNA polymerase High Fidelity (5U/μl) were assembled. For TA cloning, Platinum® Taq DNA polymerase High Fidelity was replaced with Taq DNA polymerase (5 U/μL, Fermentas, Glen Burnie Maryland, USA). The following PCR program was used: 95°C for 5 min for initial denaturation; 35 cycles using a Biometra®Tpersonal Thermocyler (Goettingen, Germany) at 94°C for 30 sec for a second denaturation process, followed by annealing temperature optimized at 4°C lesser than melting temperature of used primers; 72°C for 45 sec for DNA extension; 72°C for 5 min for final DNA extension and stored at 4°C. The amplified PCR products were separated on 0.8% agarose gel and was visualized with ethidium bromide.

Ligation reaction of vector/ DNA fragment or DNA fragment/DNA fragment

Vector ligated with fragment or fragment ligated with fragment in ratio 1:1 was carried out using T4 DNA ligase (5 U/μL). The T4 DNA ligase (5 U/μL) was obtained from Fermentas, Helsingborg, Sweden. The ligation reaction was set up in a 20 μL reaction containing 1 U of T4 DNA ligase, 2 μL of 10X T4 DNA ligase buffer and dH_2O was added up to 20 μL. The reaction was incubated at 22°C for 10 min.

Cloning and transformation of Escherichia coli DH5α

The yT&A cloning vector kit were purchased from Yeastern Biotech Co., Ltd, Shijr Taipei, Taiwan. yT&A cloning and all the ligation buffer was vortex before use. Two microlitre of the cloning vector (25 ng/μL) was mixed with 1 μL of 10X ligation buffer A and B. Three microlitre of amplified PCR products was added following with 1 μL of T4 DNA ligase and 2 μL of deionized water. This will bring up to a final volume of 10 μL. The reaction was pipetted and was incubated for 20 min at 22°C and the T4 DNA ligase enzyme was inactivated by incubating the reaction in 65°C. Two microlitre of the ligation mixture was used to transform 5×10^7 cfu/μg Escherichia coli DH5α. The transformation mixture was completely spread on LB plate containing 50 μg/mL ampicillin plus 0.1mM IPTG and 60 μg/mL X-gal. The plate was incubated 16 h at 37°C.

Colony PCR

Six colonies were randomly picked with sterilized toothpick and was dissolved in nuclease free water. The broth was added with 10 µM of each primer M13F and M13R, 10 mM dNTP mixture, 50 mM $MgSO_4$ and Taq DNA polymerase (5U/µl). The following PCR program was used: 94°C for 5 min for initial denaturation; 25 cycles using a Biometra®Tpersonal Thermocyler (Goettingen, Germany) at 94°C for 30 sec for a second denaturation process, followed by annealing temperature at 55°C for 30 sec; 72°C for 30 sec for DNA extension; 72°C for 7 min for final DNA extension and stored at 4°C. Successful colony was subjected to fermentation process and plasmid purification.

Purification of the amplified DNA(s), digested fragment(s)

Amplified DNA(s), digested fragment(s) from 1% agarose gel was excised with a sterilized scalpel in 50 mL Falcon tube. The gel slice was weighed and three volumes of Buffer QG to one volume of gel (100 mg~100 µL) were added. The tube was incubated at 50°C for 15 min and was vortexed every 3 min until the gel was dissolved completely. One gel volume of isopropanol was added and mixed gently. QIAquick spin column was placed in a 2 mL collection tube. The dissolved gel was applied to the QIAquick spin column and was centrifuged for 1 min. The flow-through was discarded. This step was repeated until all the dissolved gel solution was applied. Five hundred microlitre of buffer QG was added to QIAquick spin column and was centrifuged for 1 min. Followed this, 750 µL of buffer PE was added and was centrifuged for 1 min. The flow-through was discarded and additional centrifugation was carried out to remove all the buffers from the spin column. The QIAquick spin column was placed into a clean 1.5 mL appendorf tube and 30 µL of EB buffer was added in the center of the column to elute out the DNA from the spin column. The clean DNA was stored at -20°C.

Purification of vector(s)

Vector(s) was purified using QIAprep® Miniprep plasmid purification Kit (Qiagen, Venlo, Netherlands). The QIAprep miniprep procedure was based on the alkaline lysis of the bacterial cells followed by adsorption of DNA onto silica membrane in the presence of high salt and elution in low-salt buffer (Tris-HCl, pH 8.5). The bacterial cells were isolated from the overnight LB culture by centrifugation at high speed for 1 min and were resuspended in 250 μL of the P1 resuspension buffer (50 mM Tris-HCl, 10 mM EDTA, pH 8.0 and 100 μg/mL RNase A). The cells suspension was lysed under alkaline conditions using 250 μL of the P2 lysis buffer (200 mM NaOH and 1% SDS). The lysate was subsequently neutralised by addition of 350 μL of the P3 neutralisation buffer (3.0 M KAc, pH 3.5). Then, the mixture was centrifuged at 12 000 g for 10 min. The supernatant was transferred into a QIAprep Spin Column, centrifuged at 12 000 g for 1 min and the flow-through was discarded. The column was washed using 750 μL of the PE buffer. After centrifugation, the column was placed in a clean 1.5 mL collection tubes and elution of the plasmid DNA was carried out by addition of 50 μL of the EB buffer (10 mM Tris-HCl, pH8.5) to the column, standing for 1 min and collection of the plasmid DNA by centrifuging at 12 000 g for 1 min. The purified plasmids were kept at -20°C.

Restriction digestion and sequence analysis

The recombinant construct from white colonies was digested with HindIII and BglII (Fermentas, Glen Burnie Maryland, USA). The reaction mixture was set up in a 20 μL. One microlitre of any samples that needed to be digested was mixed with 16 μL of dH_2O, 2 μL of 10X respective restriction buffer and 2 μL restriction enzyme. If double digestion were to be carried out universal Tango™ buffer was used and 1 μL of each restriction enzyme were used. The digestion reaction was later separated using 0.8% agarose gel electrophoresis. The digestion restriction was spun down prior to incubation in optimum temperature for 1 h. The digestion reaction was then subjected to electrophoresis analysis and sequencing.

Bioinformatics analysis

Sequencing results were subjected to bioinformatic analysis. Table 2 shows alphabetically listed bioinformatics server and its functions.

Table 2: Web servers for bioinformatics' tools that were used in this study.

Sever name	Web address	Function	References
Active site prediction server	http://www.scfbio-iitd.res.in/dock/ActiveSite.jsp	predicting active site in a given protein	Singh et al., 2011
BSpred	http://zhanglab.ccmb.med.umich.edu/BSpred/	predicting binding site of proteins from amino acid sequences	Mukherjee et al., 2010
ChloroP 1.1	http://www.cbs.dtu.dk/services/ChloroP/	predicts the presence of chloroplast transit peptides (cTP) in protein sequences and the location of potential cTP cleavage sites	Emanuelsson et al., 1999
MEGA5	http://www.megasoftware.net/	molecular evolutionary genetics analysis	Tamura et al., 2011
MODWEB	http://modbase.compbio.ucsf.edu/ModWeb20-html/modweb.html	protein structure modelling	Pieper et al., 2011
NCBI	http://www.ncbi.nlm.nih.gov/	primer blast	Johnson et al., 2008
		search protein database using a protein query	
Predict Protein	https://www.predictprotein.org/	sequence analysis, secondary structure and function prediction	Rost et al. (2004); Rost 1996
ProFunc	http://www.ebi.ac.uk/thornton-srv/databases/ProFunc/	clefts and cavities in protein structure	Laskowski et al., 2005
Protein Stats	http://www.bioinformatics.org/sms2/	protein statistic result	Stothard et al., 2000
ProtParam	http://expasy.org/tools/	physico-chemical parameters of a protein sequence (amino-acid and atomic compositions, isoelectric point, molecular weight	Gasteiger et al., 2005
SDSC Biology WorkBench	http://workbench.sdsc.edu/	multiple sequence alignment	Rajesh and Manickam, 2006
		translate DNA to protein	
		predicting open reading frame (ORF)	

SignalP 4.0	http://www.cbs.dtu.dk/services/SignalP/	predicting the presence and location of signal peptide cleavage sites in amino acid sequences	Petersen et al., 2011
TMHMM 2.0	http://www.cbs.dtu.dk/services/TMHMM-2.0/	predicting transmembrane helices in protein	Sonnhammer et al., 1998

CASE STUDY 2

HPLC SCREENING TO DETERMINE TOCOCHROMANOLS COMPOSITION

INTRODUCTION

Vitamins supplies for human are only derived from diets. It is a group of compounds that are important for human health. There are basically two groups of vitamins, water soluble and fat soluble, while vitamin E falls into the fat-soluble group. The storehouse of useful vitamins and minerals are in green leafy vegetables like green *Amaranthus sp.* and *Allium porrum.*

A mixture of α -β-γ-δ- tocopherols and tocotrienols are known as tocochromanols. The different between tocopherol and tocotrienol isoforms only vary in methyl substitution on the chromanol ring and the saturation of the phytyl tail. α-tocopherol and α-tocotrienol are the most studied compounds because they have the highest biological and antioxidant activities. The levels of tocochromanols are mainly determined in dietary supply and rarely in human cells and animal cells.

Vitamin E detection using gas chromatography coupled with mass spectra leads to long, tedious and costly procedure. Several publications have been reported on other types of methods such as TLC and calorimetric. However, due to the easier applicability and the availability, HPLC technique is more frequently used as it is a well-suited technique for vitamin analysis. This study employs a HPLC method which uses a direct extraction and detection of a few microlitres samples. Such methods are suitable for a larger number of samples and are friendlier to users. Even though there are a lot of chromatographic methods available for the determination and analysis of tocochromanols, the reported method and analysis are needed to suit the current sample. Therefore, to confirm the suitability with the current sample linearity, precision, recovery and coefficient variance of repeatability and reproducibility were detected.

In this study, both α- tocopherol and α- tocotrienol was quantified after direct extraction with pyrogallol in ethanol. Pyrogallol is an antioxidant that reduces damages to vitamin E. This method was originally developed for cotyledons, but was applied on matured leaves. The sample was pre-treated using extraction solvent to break the structure of membranes, lipoproteins and liposome into which vitamin E binds. The extraction solvent will also reduce the interferences that determinate by proteins and carbohydrates that soluble in organic phase.

The aim of this study was to ensure an extraction and analytical technique that suit the determination of tocochromanols using normal phase isocratic HPLC, coupled with fluorescence detection. After validation, this method was successfully applied for the determination of the tocochromanols standards and tocochromanols in leaf extract.

MATERIALS AND METHODS

Preparation of tocochromanols standards

The standard stock solution α-, γ-, δ- tocopherols were purchased from Sigma-Aldrich Steinheim, Germany. Raw material of tocotrienols was requested from Carotech Berhad, Chemor, Perak instead of purchasing high cost tocotrienol standards.

α-δ- tocopherol (100 mg) was dissolved in 1mL of isopropanol (HPLC grade) to make a stock concentration of 0.23 M and 0.25 M, respectively. γ-tocopherol (25 mg) was dissolved in 1 mL of isopropanol to make a stock concentration of 0.03 M. For the preparation of tocotrienol standard, each standard α-γ-δ- tocotrienol (100 mg) was dissolved in 1 mL of isopropanol to make a stock concentration of 0.24 M, 0.24 M and 0.25 M, respectively. The prepared working stock was tightly wrapped in an aluminum foil and stored at -20°C to avoid volatilization of the solution. Standards were thawed prior to analysis using high performance liquid chromatography (HPLC). The solution of 10 μM of each individual and standard mixture was prepared by diluting each stock standard in an appropriate mL of isopropanol by using $M_1V_1=M_2V_2$ equation.

Serial dilution of 2 μM, 4 μM, 6 μM, 8 μM, 10 μM, 15 μM, 20 μM, 25 μM was prepared. Working solution of 2 μM, 4 μM, 6 μMand 8 μM was prepared from 10 μM stock solution and working solution of 15 μM, 20 μM, 25 μM was prepared from initial standard stock solution. A similar equation as mentioned above was used to prepare various concentrations of working solutions. The concentration was selected based on preliminary trials in order to obtain standard curve with high R^2 value. The prepared standard solutions were kept in 2 mL clear glass screw thread vials (Grace, Deerfield, USA) covered with aluminum foil and were stored at -20°C. The contents were brought to room temperature prior to analysis.

Optimisation of mobile phase for detection of tocochromonals composition

Optimised separation of each 10 μM compound in tocochromanols standard was determined by various compositions of isopropanol: n-hexane as mobile phase. The compositions of isopropanol and n-hexane were set as 0.5:99.5; 0.7:99.3; 0.8:99.2 and 1.0:99.0 with a flow rate of 1.5 mL/min. The selection for optimized mobile phase was based on the maximum separation, optimum peak height and length of run time. The mobile phase was de-gassed during the whole run using the internal degasser in the HPLC system.

Tocopherol extraction and HPLC analysis

Approximately, 500 mg of tissue was ground in liquid nitrogen and the resultant powder was transferred to 2 mL of 1% (w/v) pyrogallol (Sigma, St. Louis) dissolved in ethanol (Systerm® Chemicals, Selangor Malaysia). The extract was vortexed and centrifuged at 7500 rpm for 2 min to remove large chunks of debris. The supernatant was filtered through a 0.2 μm filter (PTFE 13 mm syringe filters) and filtrate was collected and transferred to a new 5 mL glass vial sized 7.5 cm x 2.5 cm. Extracts were stored at -20°C until further analysis.

Aliquot of filtrate (750 μL) was then dried under nitrogen gas 98-M 200 BAR max at 27°C (MOX LINDE, Selangor) and the dried samples were resuspended in 1 mL of isopropanol (Friendemann, Schmidt). The vials were covered with an aluminum foil. Extracts (20 μL) were separated on a Waters HPLC 2695 Separation Module connected with Waters 2475 Multi λ Fluorescence Detector system (Milford, MA, Ireland). The mobile phase (n-hexane:isopropanol) was pumped to a Phenomenex Luna 3u silica (2) 100A with silica guard column, at the flow rate of 1.5 mL/min. Fluorescence detection was used with excitation of 295 nm and emission of 330 nm. Unknown peaks were identified by comparing with the retention times of the tocochromanols standards.

Detection of standard tocochromanols by HPLC quantification

Tocochromanols were analysed using a HPLC system equipped with a pump (WATERS 2695-Sperations Module), a detector (WATERS 2475-Multi λ Fluorescence Detector) and an integrator integrated with EMPOWER version 1. The column, (150 x 4.6 mm) packed with silica (3 μm particles) and coupled with a guard column, was used to determine the tocochromanols separation. It was analysed under isocratic condition with 0.5: 99.5 as elution solvents with a flow rate of 1.5 mL/min. The wavelengths were set at 295 nm for excitation and at 330 nm for emission using a fluorescence detector. Integration of tocochromanols compounds was set to obtain smooth baselines and the lowest concentration at which a tocochromanols compound would produce the smallest peak on the monitor was determined from the standard curve equation. Stock solutions, filtered extracts and spiked samples each 10 μl were injected separately into the injector and was allowed to run for 15 min. The standard curve for individual tocochromanols were prepared by plotting the response (peak height) versus various concentrations (μM). Each peak in chromatogram was identified by comparing its retention time with the corresponding peak obtained from individual standard. The amount of detected tocochromanols in the extract was quantified from the standard curves. The limit of quantification (LOQ) for individual tocochromanols isomers were determined as the lowest, non-zero concentration of a given tocochromanols, selected from the range of calibration curved and expressed in μM.

Replication

Sufficient number of replicates for each treatment was included to account for variability in the test samples. Each treatment in the test series started with the same number of replicates. In this study, the experiment for each treatment was carried out in triplicates. Intra- and inter- assay instrumental precisions were determined by the analysis of the standard in one day (intra) and in different days (inter).

Statistical analysis

All results are expressed as means + SD. Coefficient variance was calculated by (SD/Mean) x 100.

CASE STUDY 3

VITAMIN E PRODUCTION BY SILENCING HOMOGENTISATE PHYTYLTRANSFERASE GENE IN GREEN *AMARANTHUS SP.* AND *ALLIUM PORRUM*

INTRODUCTION

Manipulating the plant system for vitamin production is of a great interest to plant researchers. Therefore, attempts were made to manipulate gene expression in plants. However, some of these efforts resulted in unexpected gene silencing. Gene silencing is a process within plants that moderates the activity of genes in plant tissues. It can be achieved by many ways and in this study gene silencing through co-suppression was carry out.

In co-suppression, RNAs encoded by both transgenes and homologous endogenous genes are degraded through post transcriptional gene silencing (PTGS) mechanism in plants. Post transcriptional gene silencing mechanisms will degrade any copy of the sense strand of aberrant mRNA from the transgene that has the potential to interfere with the homologous mRNA. As it degrades the aberrant RNAs, the homologous mRNA is degraded as well and this leads to gene silencing.

Homogentisate phytyltransferase gene from *Elaeis guineensis* has become the gene of choice to evaluate the co-suppression in green *Amaranthus sp.* (dicot) and *Allium porrum* (monocot) as this enzyme determines the composition of vitamin E in plant tissues. Vitamin E is a primary metabolite that consists of α-tocopherol typically found in all green vegetables and α-tocotrienol found in plant seeds. The HPT enzyme is involved in the committed step in δ and γ–tocochromanols synthesis. β and α - tocochromanols are produced by the subsequent methylation of δ and γ - tocochromanols by other groups of enzymes (MPBQ methyltransferase and γ-tocopherol methyltransferase). Suppression of a key α-tocopherol biosynthesis gene (MPBQ MT) produced mutant line of Arabidopsis with reduced levels of γ and α tocopherol but greatly increased the levels of δ and β tocopherol. Quite a number of papers reported on the over-expression of tocochromanols but very few were on tocopherol co-suppression, such as suppression of tocopherol in potato by and in Arabidopsis.

Therefore, in the present study, co-suppression strategy was employed using short cDNA that encodes the conserved region of the open reading frame of HPT gene. The partial cDNA was used in the construction of recombinant vector with Ubi1P or LHCB with/without Ubi1- intron fragment using the pDRB6b expression vectors. In addition, the transformed *Agrobacterium tumefaciens* (AGL1) was characterized based on the stability of the recombinant vector it harboured. Since most agriculture crops are not amenable to stable genetic transformation through tissue culture technique, agroinfiltration-based system was used as a tool to transform intact leaf tissues of green *Amaranthus sp.* and *Allium porrum*. Our aim was to suppress α-tocopherol and to observe the formation of α-tocotrienol.

MATERIALS AND METHODS

Plant materials and growth conditions

Elaeis guineensis, aged six weeks old, were obtained from GOLDEN HOPE, Banting, Selangor. Seeds of green *Amaranthus sp.* (8) and *Allium porrum* (717) were purchased from Leckat Corporation Sdn. Bhd. Malaysia. The number in bracket indicates the product code. Both plant seeds were sown directly in a pot containing sterilized soil mixtures containing soil, peatgro (Peatgro™, Batu Caves, Selangor)and sand in a ratio of 1:2:1 in the Transgenic Glasshouse, UPM, under natural daylight. Four to six plants were placed in each pot. Green *Amaranthus sp.* was grown for one month and *Allium porrum* was grown for four months. The plants were watered twice daily and were supplemented with fertilizer (WELGRO, Petaling Jaya, Selangor) once every two weeks. The plants were placed under complete darkness for 24 h and the water supply was stopped for two days before agroinfiltration.

Plasmids, bacterial strains, transformation and growth conditions

The plasmids and bacterial strains used for DNA manipulation are listed in Table 1. The competent cells of Escherichia coli DH5α were purchased from Yeastern Biotech Co., Ltd, Shijr Taipei, Taiwan. Escherichia coli DH5α was grown for 16 h at 37°C.

Table 1: Bacterial strains and plasmids used in this study.

Genotype/relevant properties		Source/reference
Strains		
Escherichia coli DH5α	*end*A1 *rec*A1 *rel*A1 *gyr*A96 *hsd*R17(r_K^-,m_K^+) phoA supE44 thi-1 Δ(lacZYA-argF)U169Φ80Δ(lacZ)M15F-	Yeastern Biotech Co.,
Agrobacterium tumefaciens (AGL1)	Broad host range strain for plant transformation, Rif^R	CSIRO Plant Industry
Plasmids		
pDRB1	a control vector, -linearised with *Kpn*I	CSIRO Plant Industry
pDRB5b	a donor of Ubi1-intron, control vector, -linearised with *Kpn*I	CSIRO Plant Industry
pDRB6b	an expression vector, control vector, $Spec^R$, linearised with *Kpn*I	CSIRO Plant Industry
pDRB8	a donor of Ubi1 P and NosT, control vector, -linearised with *Sma*I	CSIRO Plant Industry
pGWLSO1	a donor of leaf specific promoter (LHCB)	MPOB
yT&A	cloning vector, Amp^R	Yeastern Biotech Co.,
p5b5	recombinant vector; T-DNA region consist of Ubi1P, HPT, NosT; $Spec^R$, -linearised with *Not*I	This study
p5d9	recombinant vector; T-DNA region consist of Ubi1P, Ubi1-intron,HPT, NosT; $Spec^R$, linearised with *Not*I	This study
p5f7	recombinant vector; T-DNA region consist of LCHB, Ubi1-intron, HPT, NosT; $Spec^R$, linearised with *Not*I	This study

Agrobacterium strains (AGL1) were obtained from Dr N.M. Upadhyaya,

CSIRO, Australia. AGL1 carrying the plasmid (pDRB6b, pDRB5b, pDRB8, p5b5, p5d9 and p5f7) was streaked on LB agar (Merck, Darmstadt, Germany) plates containing 50 μg/L spectinomycin (Calbiochem® Germany) and 25 μg/L rifampicin (Calbiochem® Canada). The plates were incubated at 28°C for two days. The Agrobacteria was inoculated in 5 mL LB broth with the same antibiotics and was grown overnight at 28°C on a shaker at 150 rpm. All transformed bacterial strains were grown on Luria-Bertani (LB) agar or broth (Difco™, USA). Broth cultures were grown with shaking at 200 rpm.

Construction of the recombinant vectors

The pDRB6b vector (Figure 1) digested with HindIII/KpnI, served as the backbone vector for p5b5, p5d9 and p5f7. The Ubi1P and NosT were isolated by using HindIII/BglII and XhoI/KpnI restriction enzymes from the pDRB8 vector, respectively (Figure 2). The LHCB was isolated by amplifying the sequences using FLHCB and RLHCB primers (Table 2) from the pGWLSO1 vector (Table 1). pDRB5b vector (Figure 3) was used to isolate Ubi1- intron by using PstI/BglII. The HPT gene was amplified with FHPT2B2, RHPT2X1 and FHPT2P3 primers (Table 2) to produce two different fragment i) BglII/XhoI and ii) PstI/XhoI ends. The HPT gene having BglII/XhoI restriction sites was ligated to Ubi1P having BglII restriction sites at 3` end to NosT having XhoI restriction sites at its 5`end. The resultant plasmid was named p5b5 (Figure 2a). For p5d9/p5f7 (Figure 2b, c), the Ubi1P/LHCB was fused with Ubi1-intron fragment having PstI at 3` end prior ligating to HPT gene with PstI/XhoI ends and NosT fragment. Ligation mixture, purification of the amplification fragment mixtureand purification of vector was assembled as mention in earlier.

Verification of the recombinant construct

The recombinant and control vectors that were amplified in *Escherichia coli* DH5α were isolated as indicated in earlier section. The isolated vectors were linearized using single enzyme digestion (Table 1) as mentioned earlier. The successful recombinant vector sizes were compared with the control vectors. The positive recombinant vectors were confirmed by amplifying the HPT gene with primary and nested primers (vcFHPT2, vcRHPT2, vcHPT2F3) that flanked by Ubi1P or LHCB and

NosT fragment between LB and RB2 border (Table 2). The amplified PCR product was sequenced and analysed using the NCBI Blast.

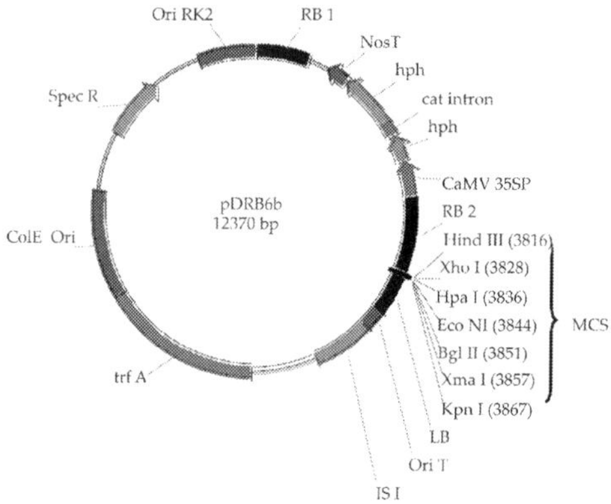

Figure 1: Physical map of control vector pDRB6b. Arrow indicates the orientation of each DNA fragments that was assembled. LB: left border of T-DNA. RB1:first right border of T-DNA. RB2: second right border of T-DNA. CAMV35SP: cauliflower mosaic virus (CaMV) promoter, hph: encodes hygromycin phosphotransferase, cat intron: intron containing catalase gene, NosT: termination sites, Spec R: spectinomycin resistant gene, Ori RK2: vegetative replication origin, ColE Ori: origin of replication, trfA: encodes replication initiation protein, MCS: multiple cloning sites with with various restriction enzyme sites (CSIRO Plant Industry).

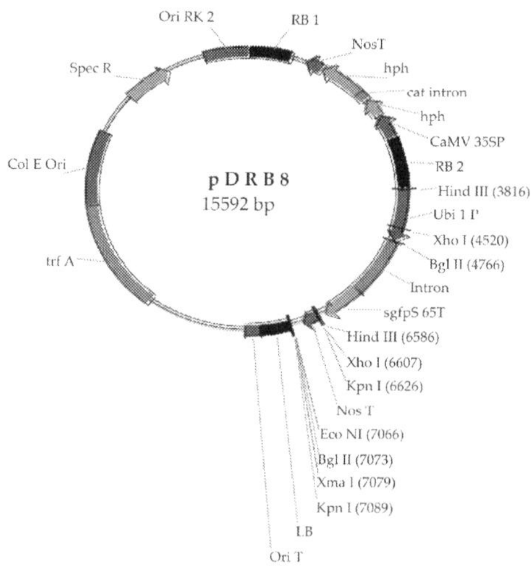

Figure 2: Physical map of control vector pDRB8. Arrow indicates the orientation of each DNA fragments that was assembled. LB: left border

of T-DNA. RB1:first right border of T-DNA. RB2: second right border of T-DNA. Ubi1P: ubiquitin promoter, Ubi1-intron: ubiquitin intron, uidA: encodes beta-glucuronidase enzyme, NosT: termination sites, CAMV35SP: cauliflower mosaic virus (CaMV) promoter, hph: encodes hygromycin phosphotransferase, cat intron: intron containing catalase, Spec R: spectinomycin resistant gene, Ori RK2: vegetative replication origin, ColE Ori: origin of replication, trfA: encodes replication initiation protein (CSIRO Plant Industry).

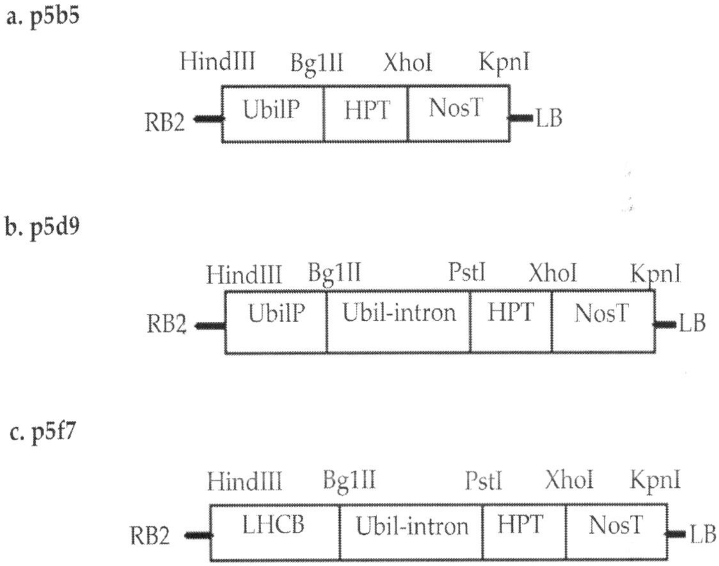

Figure 2a-2c: Chimeric constructs with homogentisate phytyltransferse (HPT) genes. RB2: second right border, LB: left border. Constructs were prepared by fusing together short sense sequences of HPT (538 bp) with ubiquitin promoter-Ubi1P (950 bp) or leaf-specific promoter–LHCB (1000 bp), ubiquitin intron (1041 bp) and NosT indicates the nos terminator (482 bp). The restriction enzyme sites in italics were used for ligations. The length of a) p5b5~ 1970 bp b) p5d9~ 3011 bp c) p5f7~ 3061 bp.

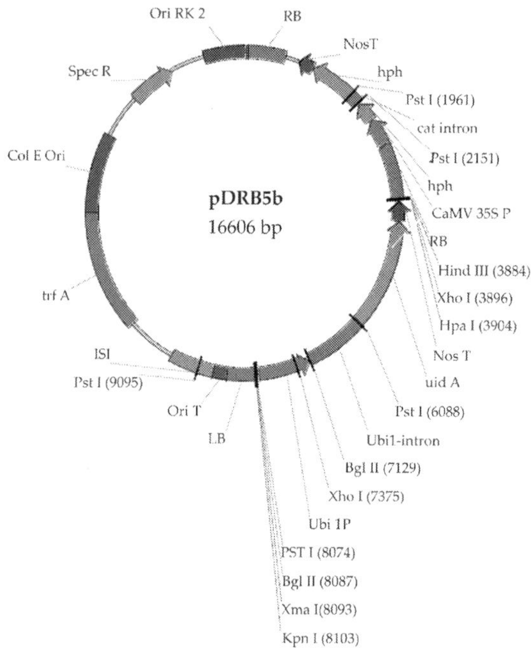

Figure 3: Physical map of control vector pDRB5b. Arrow indicates the orientation of each DNA fragments that is assembled. LB: left border of T-DNA. RB1:first right border of T-DNA. RB2: second right border of T-DNA.Ubi1P: ubiquitin promoter, Ubi1-intron: ubiquitin intron, uidA: encodes beta-glucuronidase enzyme, NosT: termination sites, CAMV35SP: cauliflower mosaic virus (CaMV) promoter, hph: encodes hygromycin phosphotransferase, cat intron: intron containing catalase gene, Spec R:spectinomycin resistant gene, Ori RK2: vegetative replication origin, ColE Ori: origin of replication, trfA: encodes replication initiation protein (CSIRO Plant Industry).

Table 2: PCR and DNA sequencing primers used in this study.

Primer name	T_m °C	Nucleotide sequnce 5′ to 3′	template
FHPT2B2	69.0	TTCAGTAGATCTATGGACGCGCTTCGCC	yT&A-HPT
RHPT2X1	69.0	TTAGGTCTCGAGGTAGGCCATCTCAAGC	yT&A-HPT
FHPT2P3	67.9	TTCAGTCTGCAGATGGACGCGCTTC	yT&A-HPT
vcFHPT2	70.6	CTACGGCCGTCCCTCCTCCC	p5b5, p5d9,p5f7
vcRHPT2	66.5	ACCGGACGCTGAAGGATCGGA	p5b5, p5d9,p5f7
vcHPT2F3	66.6	TGGACGCGCTTCGCCTACG	p5b5, p5d9,p5f7
FLHCB	63.1	AGGGGGAAGCTTAAATAAAAAGATCGG	pGWLSO1
RLHCB	63.1	TTCAATAGATCTGATGCCTTGGAG	pGWLSO1

*Sequence in italics underlined represents restriction enzyme sites

Determination of plasmid segregational stability

The recombinant plasmid (1 µg) was introduced into AGL1 strain by freeze-thawed method. The mixture was freeze in liquid nitrogen and was thawed at room temperature for 5 min. The mixture was then transferred into new 2 mL culture medium and was incubated at 28 °C for 4 h. The cells were pelleted by centrifugation for 2 min. The cell pellets were resuspended in LB medium containing spectinomycin. Hundred microlitres of cells was plated on LB agar containing 50 µg/L spectinomycin. Upon successful transformation, the plasmid segregational stability of the AGL1 was determined. A single AGL1 colony was grown in 5 mL LB broth containing 50 mg/L spectinomycin as a starter culture for two days at 28°C. Then 10 µL of the culture broth was aliquot into 5 mL of LB broth without antibiotics. Ten microlitres were removed from the overnight culture and fresh 5 mL broth was added daily until the 10th day. At the same time, an aliquot of 10 µL was taken and subjected to a serial dilution prior to plating on LB agar with and without spectinomycin (50 mg/L) to determine the colony forming units (cfu). The spectinomycin resistant colonies were recorded as a percentage of total colony forming units (cfu) per mL of culture on LB agar without spectinomycin [i.e. %spectinomycin resistant= (Spec^r cfu/ total cfu) x 100].

Preparations of Agrobacterium suspension for agroinfiltration of *green Amaranthus sp.* and *Allium porrum* leaves

Two millilitre of culture was collected in a sterile microcentrifuge by centrifugation at 8000 rpm for 10 min. The Agrobacteria pellet was resuspended in 1 mL sterilized induction medium (10.5 g/L K_2HPO_4, 4.5 g/L KH_2PO_4, 1 g/L $(NH_4)_2SO_4$, 0.5 g/L NaCitrate, 1 g/L glucose, 1 g/L fructose, 4 g/L glycerol, 1 mM $MgSO_4$, 10 mM MES, pH 5.6). The bacterial suspension was transferred to 15 mL falcon tubes. Four millilitre of sterilized infiltration medium (10 mM $MgSO_4$, 10 mM MES, pH 5.6) was added to make up 5 mL of medium supplemented with 100 μM acetosyringone (Sigma Aldrich ® Chemistry, USA) and 50 mg/L spectinomycin. Agrobacteria suspension was further grown at 28°C for 6 h. The bacterial cultures were collected by centrifugation and were resuspended in infiltration medium containing 200 μM acetosyringone.

Agroinfiltration was conducted by infiltrating agrobacteria suspension (1.0 OD) into the intercellular of the leaves that are still attached to the intact plant. The green *Amaranthus sp.* leaves were chosen for infiltration while for *Allium porrum*, all leaves were infiltrated. The leaves were injured and were submerged into the agrobacteria suspension for 1 min. After infiltration, the leaves were covered with plastic to maintain the humidity. The cover was removed after 24 h and the plant was transferred back to the glasshouse. Schematic view of the steps involved in agroinfiltration of green *Amaranthus sp.* and *Allium porrum* as shown in Figure 4a, b. Samples were harvested at day three, five and seven. Harvested samples were freeze dried and were stored at -80°C prior to analysis.

Tocopherol extraction and HPLC analysis

Tocopherol analysis was carried out as indicated earlier section.

Microbiological analysis

Pruned green *Amaranthus sp.* was further grown until the formation of wild seeds while, pruned *Allium porrum* was further grown until it reached the maturity stage. The collected seeds from green *Amaranthus sp.* and *Allium porrum* leaves were tested for occurrence of microbes as both materials derived from plants were used for agroinfiltration. The sample was sent to private laboratory to be analysed.

Statistical analysis

All experiments were repeated thrice with at least three replications for each analysis. The data presented herein are the average results of the experiments and all data were subjected to normality test. One-way analysis of variance (ANOVA) was performed for all raw data using Duncan multiple range tests. All analyses were performed at the level 5% using SPSS 17.0 software package (SPSS Inc., IL, USA).

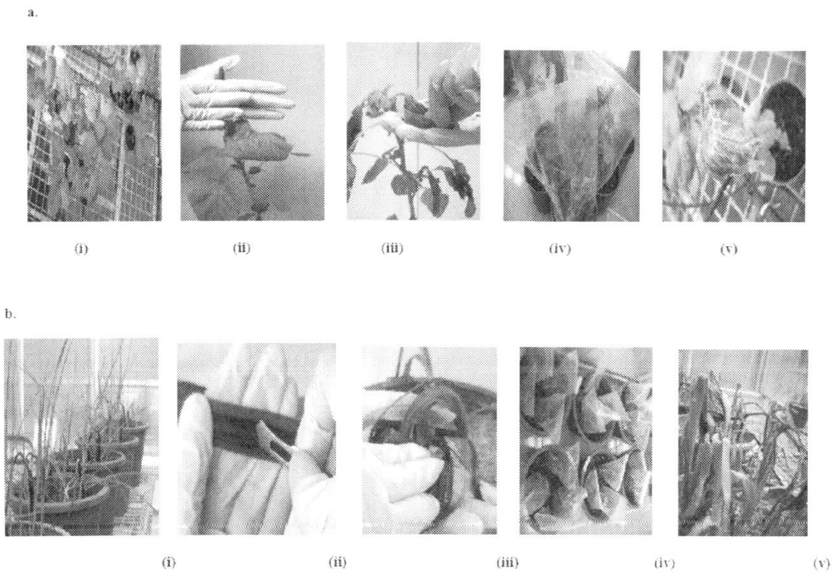

Figure 4: Schematic diagram of agroinfiltration-mediated transient transformation for (a) green *Amaranthus sp.* and (b) *Allium porrum*. Step (i) matured plant ready to be infiltrated; (ii) intercellular space was exposed (iii) bacterial suspension was infiltrated by submerging the leaves in the suspension (iv) humidity of the infiltrated plant was maintained by using plastic cover (v) after 24 h, infiltrated plant was placed at glasshouse (Transgenic Glasshouse, UPM).

CASE STUDY 4

TOCOPHEROL CYCLASE GENE SILENCING FOR VITAMIN E PRODUCTION IN GREEN *AMARANTHUS SP.* AND *ALLIUM PORRUM*

INTRODUCTION

Metabolic engineering uses a set of genetic engineering approaches to manipulate a metabolic pathway of interest in order to increase or reduce the production of useful metabolites. It requires a basic understanding of the biochemical pathways and identification of the genes and enzymes involved in the synthesis of particular metabolic compounds. Metabolic engineering usually involves an expression or a suppression of a particular gene by shifting the pathway, or by decreasing the amount of a compound, or by increasing other compounds parallelly. This modulation can result in increases in the concentration of an existing or the novel compound.

Since plants produce various ranges of metabolites, they are known as excellent sources of foods, medicines and industrial raw materials that are beneficial to human. Plant-derived vitamins (A, E, K) are of great interest due to their impact on human health. However, it is risky to rely solely on plants to produce food naturally, as plants do not provide sufficient nutrient output to meet the nutritional needs. To address these needs, attempts were made by many researchers to manipulate the gene expression in plants. However, some of their efforts resulted in unexpected gene silencing.

Gene silencing is a process in living cells that regulates the activity of genes in the tissues where they are normally functional. It can be achieved by co-suppression technologies that use post transcriptional gene silencing (PTGS) mechanisms. The PTGS mechanisms reduce copies of the sense strand of aberrant RNA. At this time, it will also degrade endogenous mRNA. This will result in the inhibition of all the mRNA that corresponds to the particular gene.

Tocopherol/tocotrienol cyclase (TC) from *Elaeis guineensis* is the gene of choice to evaluate the co-suppression mechanisms in green

Amaranthus sp. (dicot) and *Allium porrum* (monocot) because this enzyme will determine the final composition of vitamin E in any plant tissues. Vitamin E (primary metabolite) consists of α-tocopherol and α-tocotrienol, referred to as tocochromanols and is produced through two metabolic pathways: shikimate and non-mevalonate. Typically green vegetables have high level of tocopherols while plant seeds contain high level of tocotrienols.

Overexpression of TC resulted in an increase in tocopherol and a dramatic shift from γ-tocopherol to α-tocopherol. In addition, it will also led to accumulation of other antioxidant compound. Similar outcome reduction in TC led to accumulation of carbohydrates. This evidence is reasonable to show that when one of the metabolites present in the biosynthesis pathway was suppressed, the other pathway that exists will take over in synthesising other forms of the metabolites that are necessary for the organism's survival.

Co-suppression mechanism was also studied in petunia with CHS gene, in tobacco with AtFAD7 gene and in tomato with *GA 20-oxidase1* gene. In the present study, the short cDNA that encodes the open reading frame of TC gene was isolated and was expressed with Ubi1P or LHCB using the pDRB6b expression vectors. The transformed *Agrobacterium tumefaciens* (AGL1) was characterised based on the stability of the recombinant vector it harboured. Since most agriculture crops are not amenable to stable genetic transformation, agroinfiltration-based system was used as a tool for transferring foreign gene directly into intact leaf tissues of green *Amaranthus sp* and *Allium porrum*.

MATERIALS AND METHODS

Isolation of partial cDNA of TC gene

The primers for amplification were designed based on the output of the nucleotide blast analysis. These primers (FtcB2 and RtcH2) flanked the conserved region based on the pairwise sequence alignment were synthesised by scientific companies. The primer sequences are listed in Table 1. Polymerase chain reaction was carried out with 1 U/μL of Platinum® Taq DNA Polymerase High Fidelity (5 U/μL, Invitrogen™, NY, USA) in a 25 μL reaction mixture. The reaction was set up as indicated in earlier section. The following PCR program were used: 95°C for 5 min for initial denaturation; 35 cycles using a Biometra®Tpersonal Thermocyler (Goettingen, Germany) at 94°C for 30 sec for a second denaturation process, optimisation of annealing temperature to 4°C lesser than melting temperature of the primers used; 72°C for 45 sec for DNA extension; and 72°C for 5 min for final DNA extension and stored at 4°C. The amplified PCR products were separated on 0.8% agarose gel and visualised with ethidium bromide.

The T4 DNA ligase (5 U/µL) was obtained from Fermentas, Helsingborg, Sweden and the reaction was assembled. The restriction enzymes HindIII and BglII purchased from Fermentas, Vilnius, Lithuania were used in this study. The amplified DNA(s), digested fragment(s) or vector(s) were isolated using the QIAquick Gel Extraction Kit (QIAGEN Inc., CA, USA). The plasmids were prepared using the QIAprep Miniprep Kit (QIAGEN Inc., CA, USA). Restriction enzyme digestions were carried out as discussed in earlier case studies

Table 1: PCR and DNA sequencing primers used in this study.

Primer name	T_m°C	Nucleotide sequence 5` to 3`	template
FtcB2	61.7	TTAATA*AGATCT*AGGTCGAAGTATGTCC	yT&A-TC
RtcH2	59.0	AATAAATTA*AAGCTT*TTTCCCACAATTGC	yT&A-TC
vcFTC	66.5	ACAGCAGGCTGGCTTGCAGC	p4a11, p4c9
vcRTC	66.5	TTGTTGGAGCCCGCAAGGGG	p4a11, p4c9
npfTC	64.5	TGCATGGCAGCTGGCTTGTCA	p4a11, p4c9
FLHCB	63.1	AGGGGG*AAGCTT*AAATAAAAAGATCGG	pGWLSO1
RLHCB	63.1	TTCAAT*AGATCT*GATGCCTTGGAG	pGWLSO1

*Sequence in italics underlined represents restriction enzyme sites

Plasmids, bacterial strains, transformation and growth conditions

The plasmids and bacterial strains used for DNA manipulation are listed in Table 2. The competent cells of *Escherichia coli* DH5α were purchased from Yeastern Biotech Co., Ltd, Shijr Taipei, Taiwan. *Escherichia coli* DH5α was grown for 16 h at 37°C. Agrobacterium strains (AGL1) were obtained from Dr N.M. Upadhyaya, CSIRO, Australia. AGL1 carrying the plasmid (pDRB6b, pDRB5b, p4a11 and p4c9) was streaked on LB agar (Merck, Darmstadt, Germany) plates containing spectinomycin and rifampicin. The plates were incubated at 28°C for two days. The Agrobacteria was inoculated in 5 mL LB broth with the same antibiotics and was grown overnight at 28°C on a shaker at 150 rpm. All transformed bacterial strains were grown on Luria-Bertani (LB) agar or broth (Difco™, NJ, USA). Broth culture(s) were grown with shaking at 200 rpm.

Construction of the recombinant vectors

The pDRB6b vector digested with HindIII/KpnI, served as the backbone vector for p4a11 and p4c9. Ubi1P was isolated by using HindIII/BglII restriction enzyme from the pDRB8 vector (Figure 1). The LHCB was isolated by amplifying the sequences using FLHCB and RLHCB primers (Table 1) from the pGWLSO1 vector (Table 2). The pDRB8 vector was also digested to isolate NosT using HindIII/KpnI. Tocopherol cyclase gene was amplified with FtcB2/RtcH2 primers (Table 1) from yT&A-TC plasmid to produce BglII/HindIII ends. The TC gene having BglII/HindIII restriction sites were ligated to Ubi1P having BglII restriction sites at 3` end to NosT having HindIII restriction sites at its 5`end. The resultant plasmid was named p4a11 (Figure 2a). For p4c9 (Figure 2b), the LHCB was ligated to TC gene and with NosT fragment having similar restriction sites as in p4a11.

The recombinant and control vectors that were amplified in *Escherichia coli* DH5α were isolated following standard procedures. The isolated vectors were linearised using single enzyme digestion (Table 2) and the reaction was assembled as indicated earlier. The successful recombinant vector sizes were compared with the control vectors. The positive recombinant vectors were confirmed by amplifying the TC gene with primary and nested primers (vcFTC, vcRTC, npfTC) that flanked by Ubi1P or LHCB and NosT fragment between LB and RB2 border (Table 2). The amplified PCR product was sequenced and analysed using the NCBI Blast.

Table 2: Bacterial strains and plasmids used in this study.

	Genotype/relevant properties	Source/reference
Strains		
***Escherichia coli* DH5α**	*end*A1 *rec*A1 *rel*A1 *gyr*A96 *hsd*R17(r_k^-,m_k^+) phoA supE44 thi-1 Δ(lacZYA-argF)U169Φ80Δ(lacZ)M15F-	Yeastern Biotech Co.,
Agrobacterium tumefaciens	Broad host range strain for plant transformation, Rif^R	CSIRO Plant Industry **(AGL1)**
Plasmids		
pDRB5b	a donor of Ubi1-intron,*uid*A, linearised with *Kpn*I	CSIRO Plant Industry
pDRB6b	an expression vector,Spec^R, linearised with *Kpn*I	CSIRO Plant Industry
pDRB8	a donor of Ubi1 P and NosT, linearised with *Sma*I	CSIRO Plant Industry
pGWLSO1	a donor of leaf specific promoter (LHCB)	MPOB
yT&A	cloning vector, Amp^R	Yeastern Biotech Co.,
yT&A-TC	recombinant vector, a donor of TC gene	This study
p4a11	recombinant vector; T-DNA region consist of Ubi1P, TC, NosT; Spec^R, linearised with *Not*I	This study
p4c9	recombinant vector; T-DNA region consist of LCHB,TC, NosT; Spec^R, linearised with *Not*I	This study

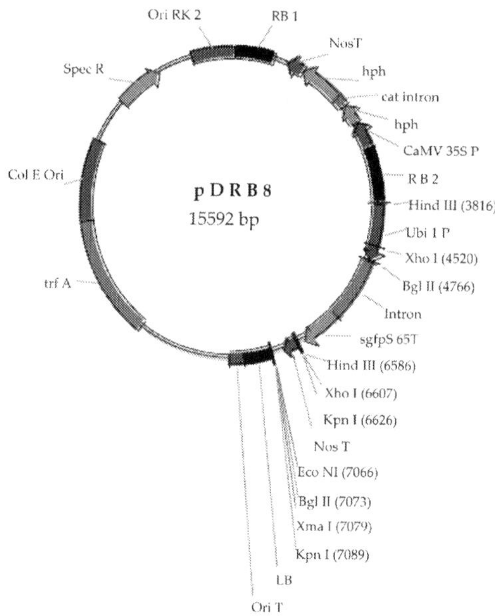

Figure 1: Physical map of control vector pDRB8. Arrow indicates the orientation of each DNA fragments that was assembled. TrfA :encodes essential replication initiation protein, ColE Ori: origin of replication from plasmid pBR322, OriRK2: origin of replication in *Escherichia coli* and *A. Tumefaciens,* oriT: origin of transfer, LB: left border of T-DNA. RB1:first right border of T-DNA. RB2: second right border of T-DNA.Ubi1P: ubiquitin promoter, Spec R:Spectinomycin resistance cassette, hph:encodes hygromycin phosphotransferase enzyme, sgfpS65T: encodes green fluorescent protein, NosT: termination sites, CAMV35SP: cauliflower mosaic virus (CaMV) promoter, Ubi1-intron:

ubiquitin intron, cat intron: intron containing catalase gene (CSIRO Plant Industry).

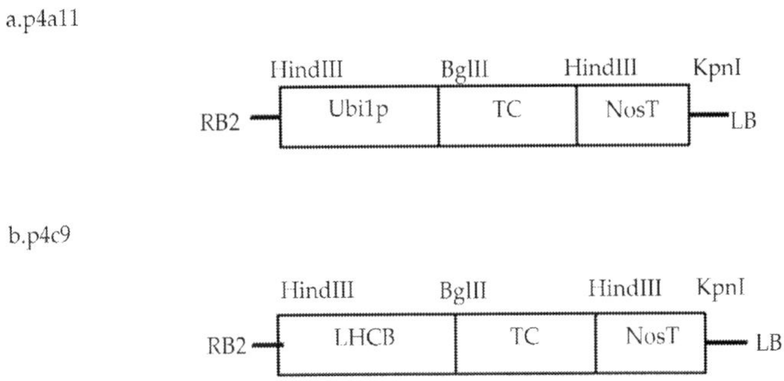

Figure 2: Chimeric constructs with tocopherol cyclase (TC) genes. RB2: second right border, LB: left border. Constructs were prepared by fusing together partial sense sequences for TC (622 bp) with maize ubiquitin promoter-Ubi1P (950 bp) or leaf-specific promoter–LHCB (1000 bp), NosT indicates the nos terminator (503 bp), the restriction enzyme sites in italics were used for ligations, the length of a) p4a11~2075 bp and b) p4c9~2125 bp.

Evaluation of transient expression

Pieces of leaves from green *Amaranthus sp./Allium porrum* that were injected with pDRB5b were assayed for GUS analysis. This was done randomly to tract the time point of GUS expression driven by Ubi1P. The GUS staining was performed. The infiltrated leaves were soaked in the GUS staining buffer (5 mg of X-Gluc in 0.1 mL of N,N`-dimethylformamide in 10 mL of buffer containing 50 mM phosphate buffer, 15 mM EDTA, 0.01% Tween-20 pH 7.0). They were incubated for 16 h at 37°C. The stained leaf tissues were rinsed with 70% ethanol until complete decolourisation. Pieces of leaves from plant injected with pDRB6b (empty vector) served as control.

CASE STUDY 5

FEMALE REPRODUCTIVE ORGANS AS THE TARGET OF AGROBACTERIUM-MEDIATED TRANSFORMATION IN GREEN *AMARANTHUS SP.*

INTRODUCTION

Crop plant produces a range of primary and secondary metabolites. Due to the limitation in genetic resources, plant breeding may require a major investment in time and space before the desired trait is obtained. Thus, genetic transformation of crop has become an important alternative in plant breeding program.

Delivery and expression of foreign genes in plants are carried out through stable genetic transformation and transient transformation. Agrobacterium-mediated transformation offers advantages in transferring copies of genes of interest by using either tissue infiltration or floral dip. This method is applied in preference to tissue culture-based techniques as it directly produces transformed seed circumventing and avoiding the lengthy time and somaclonal variation.

Green *Amaranthus sp.* is one of the most important leafy crops which can be found throughout the tropics and in many warm temperate regions. Studies have been conducted to develop technologies that aim in exploiting green *Amaranthus sp.* due to its high nutritional value. Stable Agrobacterium-mediated floral dip transformation has been developed and successfully applied in the popular model plant, *Arabidopsis thaliana* and the model plant for legumes, *Medicago truncatula.* Since both green *Amaranthus sp.* and Arabidopsis are both dicot plants with bisexual flowers, it is understandable that they share a common ancestor and similar anatomy therefore amenable to the same transformation technique (Figure 1).

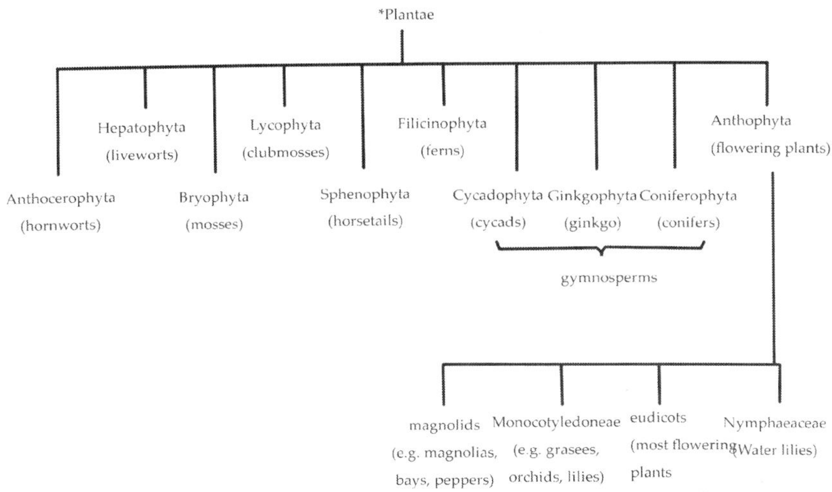

Figure 1: Simplified classification of land plants. Anthophyta consists of Angiospermophyta (Monocotyledoneae and Eudicots also known as dicotyledoneae) and Magnoliophyta (Magnolids and Nymphaeceae). The angiosperms are dominant plant and show the most advanced structural organisation in the plant kingdom. Arabidopsis thaliana and green *Amaranthus sp.* fall within the dicotyledoneae group with bisexual flowers.

Successful production of transgenic plants are also heavily dependent on a suitable selection technique to screen for transformants. There are few selections of antibiotic resistance genes such as hygromycin phosphotransferase conferring hygromycin B resistance (hph^r), bacterial neomycin phosphotransferase (nptII) conferring kanamycin resistance and Streptomyces hygroscopius phosphinothricin acetyltransferase (bar) conferring phosphinotricin (PPT) basta resistance. The effectiveness and ease of use influence the choice of suitable selectable marker gene. A simple leaf spraying method developed for selection using Basta which had enabled large scale screening of plantlets, but in the present study, BASTA was replaced with hygromycin selection.

Transformation and screening of green *Amaranthus sp.* was carried out based on the reported method. The hygromycin resistant gene was driven by the constitutive cauliflower mosaic virus 35S promoter which was located within the recombinant vector cassette carrying a key vitamin E biosynthetic gene. The HPT gene was driven by a constitutive ubiquitin promoter from maize (Ubi1P) or a leaf-specific promoter from oil palm (LHCB), with or without an intron fragment. The success in the transformation and screening of green *Amaranthus sp.* floral would widen the possibilities for the exploitation of the available reported method on many other alternative crops. However, this was not applied on *Allium porrum* as it is difficult to generate inflorescences in this plant.

MATERIALS AND METHODS

Vectors

The pDRB6b vector digested with HindIII/KpnI served as the backbone vector for p5b5, p5d9, p5f7, p4a11 and p4c9 recombinant vectors. The recombinant vector was constructed. The constructs were transformed separately into AGL1 strain using freeze-thaw method.

Floral transformation in green *Amaranthus sp.*

Seeds were sown in a pot containing soil, peatgro (Peatgro™, Batu Caves, Selangor)and sand in a ratio of 1:2:1 in the Transgenic Glasshouse, UPM, under natural daylight. The plants were allowed to grow into maturity until the length of the inflorescences were in the range of between 4 to 6 cm. *Agrobacterium tumefaciens* strain (AGL1) carrying the recombinant vectors from the glycerol stocks were streaked onto LB agar containing 25 mg/L rifampicin and 50 mg/L spectinomycin and was incubated at 28°C for two to three days in the dark. A loop full of colonies was chosen to inoculate 5 mL LB liquid medium with antibiotics in a 50 mL capacity Falcon tube and was incubated for two to three days on an orbital shaker at 180 rpm. The optical density (OD) of the bacterial culture was measured by spectrophotometerand OD_{600} of 0.7-1.0 was sufficient for the transformation event. Two millilitre of culture was collected in a sterile microcentrifuge and was centrifuged at 8000 rpm for 10 min. The agrobacteria pellet was resuspended in 1 mL sterilized induction medium (10.5 g/L K_2HPO_4, 4.5 g/L KH_2PO_4, 1 g/L $(NH_4)_2SO_4$, 0.5 g/L NaCitrate, 1 g/L glucose, 1 g/L fructose, 4 g/L glycerol, 1 mM $MgSO_4$, 10 mM MES, pH 5.6). The bacterial suspension was transferred to 15 mL Falcon tubes. Four millilitre of sterilized infiltration medium (10 mM $MgSO_4$, 10 mM MES, pH 5.6) was added to make up 5 mL of medium, supplemented with 100 µM acetosyringone and 50 mg/L spectinomycin. Agrobacteria suspension was further grown at 28°C for 6 h. The bacterial cultures were collected by centrifugation and were resuspended in infiltration medium containing 200 μM acetosyringoneand Silwet L-77 was replaced with 0.01% (v/v) Tween 20 (Sigma®, USA).

Inflorescences were inoculated by performing direct drop-by-drop inoculation to every flower by using micropipette. Three additional inoculations were applied every four days to inoculate most of the flowers in each plant. The plants were injected in a shaded area away from direct sunlight and were allowed to remain at similar condition overnight. The treated plants were placed back to the green house and allowed to grow normally. Watering of the plants were continued until the siliques started to lose their green colour and until 50% of the plant became yellow.

Seeds were harvested once the siliques were notably brown by cutting the whole plant and were left to dry in the oven at 40°C. Once dried, the seeds were separated from the bolts, leaves and stems. The seeds were stored at 4°C for further analyses. The procedure was summarised in Figure 2.

Figure 2: Overview of floral dip transformation of green *Amaranthus sp.*

a) inflorescence ready to be inoculated with AGL1 carrying p5b5, p5d9, p5f7, p4a11 and p4c9 (trangenic constructs)

b) drop-by-drop method was applied

c) inflorescence was left to become mature until siliques turn brown

d) plant materials undergo drying process

e) sowing to separate the seeds from dried plant material

f) harvested green *Amaranthus sp.* Seeds

Seed yield was determined to ensure that the drop-by-drop technique in floral dip transformation was not affecting the seed productivity. Only black seeds were sorted for this purposes.

Expression studies and transgene analysis using the hph selectable marker genes

Seeds from floral dipped plants were sown into soil. Twenty days after sowing, the young seedlings were sprayed with 150 mg/L hygromycin containing 0.01% (v/v) Tween 20 daily for five days.

Young seedlings that showed no necrosis on the leaves after the 5th day were considered to contain a functional hph gene. Hygromycin resistant plants were transferred to new pots and allowed to grow bigger. To further confirm the transgenicity of the hph resistant plants, leaf pieces of one month old plants were excised and analysed by hph histochemical assay.

To detect the expression of the hph gene, leaf disk assay in hygromycin solution was conducted to further distinguish the transformed plants. Leaf samples from young plants were placed in a 24-well microtitre plate containing 0.5 mL of 150 mg/L hygromycin. Three leaf disks from a non-transformed green *Amaranthus sp.* plant at a similar growth stage were used as a negative control. The leaf disks were placed under light at 25°C for five days. The leaf disk resistance to hygromycin remained green, indicating that the hph gene was successfully integrated into the genome of the transgenic green *Amaranthus sp.*

Fresh plant leaf materials were obtained by using FLOTING PUNCH (adoro^{TMA}). The leaf sample was crushed with 100 µL pipette tip by pressing against the microcentrifuge wall until there were no large pieces of tissues left. Five hundred microlitre of dilution buffer (0.5 N NaOH) was added to every disk according to $3(\pi/j^2)$; j^2=d=4.5 mm. (π= 22/7; J^2=d= diameter). The supernatant was centrifuged using Eppendorf Centrifuge 5810R and 0.5 µL of fresh plant tissue solution was used for preparation of a 20 µL Phire® Plant Direct PCR reaction (Finzzymes, Finland). The PCR reaction consist of 7.6 µL of distilled water, 10 µL 2x Phire® Plant PCR buffer, 1 µL of each 0.5 µM primer HyG (F:5`TCCGGATGCCTCCGCTCGAA3`) and primer HyG (R:5`ATGCAGCTCTCGGAGGGCGA3`) and 0.4 µL Phire Hot Start II DNA polymerase (F-130X). The cycling reaction was set as follows: Initial denaturation at 98°C for 5 min, second denaturation at 98°C for 45 sec, annealing at 72°C for 45 sec, extension at 72°C for 40 sec and final extension was at 72°C for 1 min, the cycle was hold at 4°C. The

PCR product was analysis on 0.8% agarose gel and was sequenced. The sequencing result was analysed using nucleotide blast NCBI.

Statistical analysis

Bacterial growth, transformation event and percentage of seed productivity data were expressed as mean values accompanied by standards error from triplicate reading performed repeatedly.

CASE STUDY 6

PRIMER DESIGN FOR PCR AND REAL TIME ANALYSIS IN PLANTS

Primer design for PCR and Real Time analysis will be carried out using free on-line software. The rationale of the process will be discussed in class. Banana (*Musa acuminata*) genes will be used as the model crop in this practical.

Primer design for conventional PCR

Instruction:

1. Carry out a literature review on Banana Housekeeping genes/reference gene and select the gene of choice.
2. Go to the NCBI website

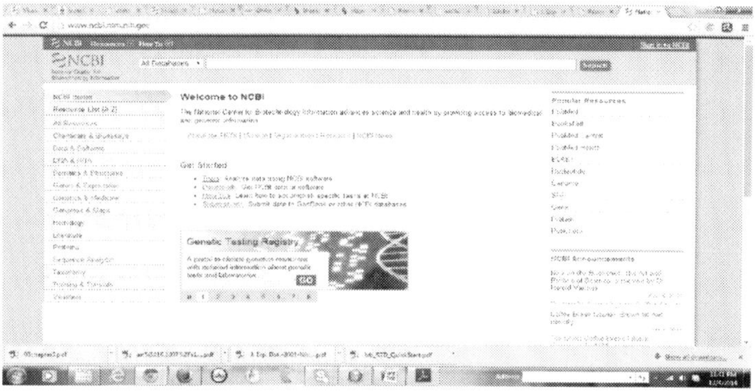

3. Locate the respective gene sequence (under Nucleotide link).
4. Go to the BLAST link.
5. Choose the software Primer-BLAST.
6. Paste your sequence into the sequence box.
7. Pick the primers generated.
8. Choose the best primer sequence.

If you need to synthesise these primers you may do so through commercial entities. One such company is given here as an example. You may locate your own scientific company at your location.

Go to FirstBase websites (http://www.base-asia.com/) to order primers according to the website instructions.

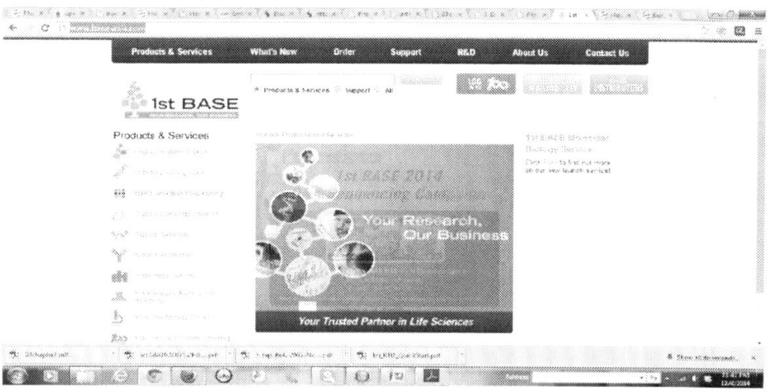

Primer design for real time

Instruction:

- Literature review on Banana Housekeeping gene/reference gene.
- Go to NCBI website.

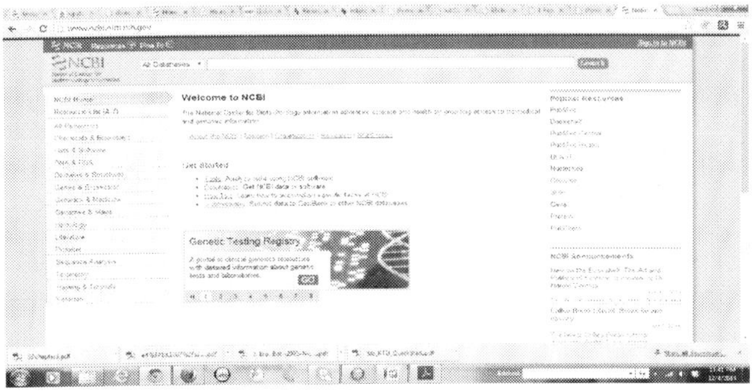

- Locate the respective gene sequence (under Nucleotide link).
- Go to the Primer3 website at: http://frodo. wi.mit.edu/primer3/

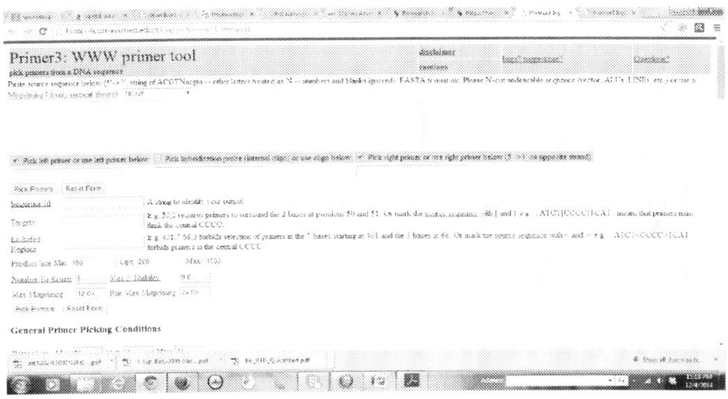

- Paste located gene sequence in Primer3 sequence box.
- Repeat steps 7-10 if necessary.

Exercise

1. What is the difference between primers for normal vs Real time PCR?
2. List the characteristics of a good PCR primer sequence and how you have selected the optimal sequences for your gene of choice

CASE STUDY 7

RNA AND DNA ISOLATION AND ANALYSIS

RNA and DNA extraction from Banana leaves

Both RNA and DNA can be extracted from Banana leaves efficiently using conventional method. The method below was optimised for the isolation of both RNA and DNA isolation from plant tissue samples that contain high phenolic compounds and polysaccharides.

Consumables and Equipment

Pipettes (1 ml, 200 µl, 100 µl, 10 µl), sterile tips (blue, white and yellow), sterile 2 ml and 1.5 ml microcentrifuge tubes, gloves, Microcentrifuge, Heating water bath, small liquid nitrogen/Dewar, Mortars and pestles, sterile spatula, Liquid Nitrogen, Agarose gel, Gel Electrophoresis system

Reagent needed

CTAB Buffer (RNA and DNA)- 2% CTAB ;20 mM EDTA (40 ml EDTA stock (0.5M));100 mM Tris-Cl pH 8.0 (100ml Tris-Cl stock (1M)); 1.4M NaCl (280 ml NaCl stock (5M)) make up to 1 Litre with water, pH 7.5 - 8.0and autoclave+ 0.2% Mercaptoethanol –prior use); β-mercaptoethanol, Phenol, Chloroform, Isoamylalcohol, Cold Isopropanol, 76% Ethanol 10 mM sodium acetate TBE buffer, Ladder, 6X loading dye, nuclease free water

Precautions

Wear clean gloves to avoid contamination. Handle liquid nitrogen with extra care. Handle all solvents in fume hood. Ultraviolet light can damage eyes. Never look at unshielded UV lightsource.

Instructions 1: RNA Isolation

1. Harvest tissue of interest from field.
2. Drop it in liquid nitrogen.
3. Grind tissue with sterilize pestle and mortar.
4. Carry out extraction using sterilize Extraction Buffer.
5. Add 2% β-mercaptoethanol.
6. Incubate 15 minutes at 65°C.
7. Add equal volume of P: C: I follow by vortexing.
8. Centrifuge at 12 000 rpm for 15 minutes.
9. Carefully transfer the upper layer into a fresh tube and add equal volume of P: C: I.
10. Centrifuge at 12 000 rpm for 15 minutes.
11. Carefully transfer the upper layer into a fresh tube and add equal volume of C: I
12. Centrifuge at 12 000 rpm for 15 minutes.

13. Carefully transfer the upper layer into a fresh tube and add equal volume of isopropanol.
14. Mix and placed at -20°C.
15. Centifuged at 12 00 rpm for 15 minutes. Decant the supernatant and dry the pellet.
16. Dissolve the pellet in nuclease free water.
17. Stored dissolved RNA at -80 °C.

Instructions 2: Genomic DNA Isolation

1. Preheat 1ml CTAB with 2% mercaptoethanol at 60-65 °C.
2. Grind frozen leaf with liquid nitrogen.
3. Scrape powder into dry tube using sterilize spatula and add pre-heated buffer and mix gently. Adjust CTAB volume to give a slurry-like consistency, mix occasionally.
4. Incubate for 10 min at 60 ° C.
5. Add equal volume of chloroform/iso-amyl alcohol (24:1) and mix for about 3 min.
6. Transfer the slurry into 2 ml microcentrifuge tube.
7. Centrifuge at 10 000 rpm for 10 min.
8. Remove the supernatant to clean tube, repeat chloroform extraction once. Supernatant should be clear without debris.
9. Precipitate DNA with 0.66 vol. of cold isopropanol.
10. Pellet down DNA by centrifuging for 2 min at 10 000 rpm.
11. Add 1 μl 10 mg/ml RNAse to each 1 ml T.E./DNA mixture and incubate for 60 min at 37 °C.
12. Dilute with 2 volumes TE and add 0.3vol 3M Sodium acetate (pH 8) + 2.5 vol cold 100% ethanol.

13. Spool DNA out. Air dry and dissolved the pellet in nuclease free water.
14. Stored dissolved DNA at 4 °C.

Instructions 3: RNA and DNA Analysis

1. Prepare 1% agarose gel with GelStain.
2. Stored samples were thawed in ice.
3. Mix 6 µl of sample with 2 µl of 6X loading dye.
4. Load 8 µl of samples into each well.
5. Run gel electrophoresis at 100V at 35 min using TBE buffer.
6. View gel using the GelDoc system.

Questions:

1. How would you prepare 500 ml of 10M Tris-HCL (pH 8.0); 250 ml of 10% SDS; Choloroform/isoamlyalcohol (24:1).
2. List down all the functions of the used solvent and buffers.
3. Why is RNAs are more easily degraded than DNA. What makes DNA more stable?
4. Describe the principles of the A260/A280 and A260/A230 ratios.
5. Do you think that a similar protocol that is as applied here can be used for other plant material such as root, tuber or for other plant species? Please explain you answer.

CASE STUDY 8

POLYMERASE CHAIN REACTION

Consumables and Equipment

Pipettes, sterilize tips, microcentrifuge, racks, fine tipped markers

Reagents

5 X Go Taq Flexi; 25 mM MgC12; Go Taq Flexi; 10 mM dNTP mix; Primer (100μM); Template; nuclease free water

PCR setup Protocols

5X Green or Colorless		
GoTaq® Flexi Buffer	10μl	1X
MgC12 Solution, 25mM	2–8μl	1.0–4.0mM
PCR Nucleotide Mix, 10mM each	1μl	0.2mM each dNTP
upstream primer	Xμl	0.1–1.0μM
downstream primer	Yμl	0.1–1.0μM
GoTaq® DNA Polymerase (5u/μl)	0.25μl	1.25u
template DNA	Zμl	<0.5μg/50μl
Nuclease-Free Water	to 50μl	

PCR thermal cycling procedure:

Step	Temperature	Time	Number of Cycles
Initial Denaturation	95°C	2 minute	1 cycle
Denaturation	95°C	0.5–1 minute	
Annealing	42–65°C	0.5–1 minute	
Extension	72°C	1min/kb	25–35 cycles
Final Extension	72°C	5 minutes 1 cycle	
Incubate	4°C	infinite	

Questions

- How would you prepare Ethidium Bromide stain (EtBr)? Describe the procedure of staining.
- What are the safety precautions that are necessary during handling of EtBr?

CASE STUDY 9

REAL-TIME ANALYSIS

Consumables and Equipment

Pipettes, sterilize tips, microcentrifuge, racks, fine tipped markers; MicroAmp® Optical 8-Tube Strip (0.2 ml)

Reagents

cDNA synthesis kit, Primers, Real-Time buffers

Instruction 1: Amplification of cDNA

Refer to Appendix

Instruction 2: Assembly of reactions

Thaw primers, reagent and buffers at RT and cDNA samples on ice.

Design a real-time RT-PCR assay using MicroAmp® Optical 8-Tube Strip (0.2 ml). We recommend a triplicate (3 wells) for each sample.

Carefully calculate the amount of volume of primers, reagents, buffers and nuclease free water needed. Assemble a reaction mixture for each gene of interest in a microcentrifuge tube.

Accurately aliquot the same volume of the reaction cocktail to each well accordingly.

Spin down the microcentrifuge prior placing it the system.

Instruction 3: Programming and performance of ABI 7500

Instruction 4: Analysis of Data with ABI 7500

Question;

List down the errors that are commonly faced during setting up the real-time PCR analysis.

Describe the latest technology in Real-Time analysis.

CASE STUDY 10

DATA ANALYSIS FOR REAL TIME ANALYSIS

In 2001, Pfaffl described a method for expressing the ratio of a target gene transcript relative to a control, calculated using solely PCR amplification efficiencies and crossing point (CT) differences. It only provides relative changes. One significant problem with relative quantification is that it is sometimes difficult to find a suitable housekeeping gene whose expression is not affected by the treatment. This can result in erroneous conclusions about the extent of gene expression changes.

1. EP refers to efficiency as defined by Pfaffl and is determined for the primer set that amplifies the target sequence (EP-target) and the primer set that amplifies the housekeeping sequence (EP-housekeeping).

2. The term CT -target refers to the difference in CT values between the control and treatment (CTcontrol − CTtreatment) for the target sequence and the term CT -housekeeping refers to the difference in CT values between the control and treatment (CT control −CT treatment) for the housekeeping sequence.

3. To calculate a relative quantification comparison, compare the CT for the target sequence among the various treatments (CT control − CT experimental).

4. Calculate the amplification efficiency values for each of your primer pairs (expressed as a value between 0 and 1). Convert each amplification efficiency measure into Pfaffl efficiency (EP) measures by adding 1.0 (e.g., 0.9 + 1.0 = 1.9).

The reasoning behind this conversion is that an efficiency of 100% (or 1.0) is equal to a 2.0-fold increase per cycle. An efficiency of 90% (0.9) is equal to a 1.9-fold increase per cycle.

5. Calculate the fold change in expression between the two samples (e.g., control versus experimental) using $(Ep)=\Delta^{Ct}$

6. To normalize the two samples using the housekeeping gene, simply divide the fold change for the target sequence by the fold change for the housekeeping gene (ideally 1.0 or very close to 1.0), expressed as (Ep)=Δ^{Ct} target/Δ^{Ct} housekeeping.

7. To calculate fold changes within the housekeeping gene between to samples (Ep)=Δ^{Ct} housekeeping root/Δ^{Ct} housekeeping leaf.

CASE STUDY 11

PROPOSED METHOD FOR MOLECULAR DETECTION OF *FUSARIUM OXYSPORUM* F. SP. *CUBENSE* RACE 4 IN INFECTED CAVENDISH

INTRODUCTION

Banana fruit is originated from India, they are the world's biggest producer. In Malaysia, banana remains the second most important fruit crop. About 265ha of land was used to grow banana and the number are getting down due to the outbreak of diseases. The most devastating disease is Fusarium wilt caused by *Fusarium oxysporum* f. sp. *cubense* (Foc race 4). The genus of Fusarium often capable to modified its biological characteristics to adapt with the environment. This range of adaption will creates a wide variety of bioactive secondary metabolites that are often become toxic to plant.

Fusarium which is a soil borne fungus colonizes into the vascular system of the host plant through root and blocks the movement of water to the upper part of the host plant. There are no cures or remedies compared to the rate of the diseases spreads to other parts of the country. Presently, eight races of the pathogen (race 0, 1A, 1B/C, 2,3,4,5 and 6) was identified and Race 4 mainly will infect Cavendish. All Foc when in contact, infects the lateral roots of banana plant will shows symptoms of necrosis and rotting. This symptom will lead to dying of the infected banana plant. In Malaysia, the outbreaks started and initially happened in Johor, but now it has started to spread to other states including Perak and Selangor. But this not ends here, as the pathogen will continue to survive in soil for long period of time before infecting other banana plants. Therefore, there is a urgent need to develop an assay methods for rapid molecular detection of *Fusarium oxysporum* f. sp. *cubense* (Foc race 4).

Gene expression analysis is extremely important in many fields of biological research. Understanding the expression pattern of genes will provides a useful mean of studying the complex regulatory networks occurring in living organisms. Molecular methods such as polymerase chain reaction (PCR) and real-time PCR are among the widely used methods to measure the levels of gene detection and expression. Polymerase chain reaction has been developed to support and replace morphology based study in identifying microorganisms. This assay is commonly developed for differentiating plant pathogenic isolates. While, in qPCR, it provides a rapid mean, relatively simple and coupled with a high level of sensitivity towards simultaneous measurement of gene expression across different sample. For accurate and reliable analysis of target gene expression, normalization is essential to correct the non-specific variations because of the differences in amount of template used and its quality that can affect the efficiency of the qPCR reactions. Normalization also allows the direct comparison of normalised transcript expression levels between samples and reference gene.

In this study, users are going to analyse the interaction of the Foc with the plant tissue as it is an important section in disease management. To support this objective, real time analysis of the candidate gene during infection and DNA detection analysis in infected plant tissues will be carry out.

MATERIALS AND METHODS

Plant Material

Infected root sample.

Growth of fungal

Fusarium oxysporum f. sp. *cubense* race 4 wilt pathogens from infected soil will be grown on PDA agar and broth at 28 °C.

DNA isolation

Dried fungal mycelium (1g) overnight grown culture (1.5ml) will be frozen instantly in liquid nitrogen and will be ground to fine powder using mortar and pestle. Genomic DNA will be extracted using CTAB method and the DNA samples will be dissolved in 1 x TBE buffer for further analysis. For DNA extraction, protocol from Sambrook et al. (1989) will be adapted.

RNA isolation

The extraction of RNA from infected and non-infected root samples will be carried out using CTAB method. 500 μl extraction buffer will be pre-warmed to 65°C in a water bath by adding 20 μl β-mercaptoethanol. 0.1-0.2 g of tissues will be ground in mortar using liquid nitrogen. The frozen powder will be quickly transferred to the pre-warmed extraction buffer and will be mixed completely by inverting the tube. Then, it will be incubated at 65°C for 10 minutes with vigorous shaking for several times. An equal volume at 10 000 g for 10 minutes at 4°C. The very viscous supernatant will be transferred to a new tube and re-extracted with an equal volume of chloroform:isoamylalcohol. Then, it will be centrifuged again as above. The supernatant will be collected very slowly and carefully to avoid taking the cell lysates in the chloroform layer. The supernatant will be centrifuged at 30 000 g for 20 minutes at 4 °C to precipitate the pellet and the insoluble material will be discarded. After that, 0.25 vol of 8M LiCl will be added to the supernatant, mixed welland stored at 4°C. The viscous supernatant will be completely discarded and the pellet will be washed with 75% ethanol 3 times to remove the remaining and will be air dried in for 10 minutes. The RNA will be dissolved in DEPC-treated water and stored.

First strand cDNA Synthesis

First strand cDNA synthesis will be synthesized by transcribing 500 ng of total RNA with Transcript One-Step gDNA Removal cDNA synthesis SuperMix from TranGen Biotech Co. Ltd. The reaction will be carried out in 20 μl reaction using random hexamers according to manufactures instruction. The concentration of cDNA will be measured and 20 ng will be used in real time qPCR reaction. All cDNA will be stored at -20 °C until further usage.

Quality control DNA and RNA

The quantity and quality of the extracted DNA and RNA sample will be checked using Nano spectrophotometer. Meanwhile, the integrity of RNA and DNA samples will be also checked by agarose gel electrophoresis. For each method, the measurement will be done with triplicate biological samples. All RNA and DNA samples will be adjusted to the same concentration.

Primer design and PCR amplification

The primer set to detect Foc race will be designed and designated as FOC F (forward primer) and FOC R (reverse primer). This primerswill be used to detect Foc race 4 in root infected with *Fusarium oxysporum* f. sp. *cubense* race 4 sampled on day 2, 4 and 6. The PCR amplification will be carried out according to manufactures recommendation. Real time primers were designed by using Applied Biosystem software. Total of 4 primers will be used to detect the gene of interest which identified as PR- pathogenesis-related protein 1, PAL-phenylalanine ammonia lyase, CAT- catalase 2 and XYL-partial xylanase inhibitor. Normalization will be carried out using 2 set of primers UBQ- Ubiquitin and Ribosomal protein-RPS.

Sensitivity of FOC primers

For the sensitivity experiment, a 20 µl PCR mixture containing10 to 100 ng DNA, 1x reaction buffer Phusion Flash PCR Master Mix, 0.5 mM of each specific primer will be prepared. The parameters for PCR cycle will be according to manufactures instruction. PCR product will be subjected to electrophoresis in 1.5 % agarose gel.

Real time qRT-PCR Assays

Real-time PCR will be performed in 8 well strip tubes with 7500 Real Time PCR system ABI SDS 7500 machine (Applied Biosystems) and universal cycling time will be applied (95°C for10 min, 40 cycles of 15 sec s at 95°C and 60°C for 60 s) in a final volume of 20 µl. The prepared reactions will contain SYBR Green Master Mix (Kappa Biosystem). 10 µM forward and reverse primers of a gene specific, 2 µl of the diluted cDNA template and 2 µl RNase free water will be included in the reaction setup. A no template control (NTC) will be included in each run for each gene. Each experiment will be conducted in three technical replicates with at least 3 biological replicates. To check for the specificity of PCR amplification, dissociation curve will be generated. The Ct values will be automatically determined for each reaction.

Statistical analysis

All data will be statistically analysed using available tools.

CASE STUDY 12

PROPOSED METHOD FOR BIOINFORMATIC ANALYSIS

Musa acuminata is a globally popular fruit produced in a large scale in.Asia & Africa. In Asia, India is the largest and greatest producer of *Musa acuminata* which delivering 29.8 million tons yearly followed by Uganda, China and Philippines. Although Cavendish remains the global export variety, cultivation of banana in different regions for local consumption differs widely. *Musa acuminata* cv Berangan is native to the Asian tropics including Malaysia Indonesia, Philippines, Australiaand also East Africa. In Malaysia it is a popular cultivar consumed as a dessert. This cultivar may be able to complement the globally exported banana such as Cavendish as it consist of an acceptable level of acidity, slightly dry starchy texture, good flavour and a reasonable shelf life compared to other local varieties.

However, this cultivar is also under attack from wilting diseases similar to Cavendish, caused by the soil-borne fungus *Fusarium oxysporum* f. sp. *cubense* (Foc). Foc is a complex pathogen manifesting as subtypes or races that have resulted in outbreaks in Latin America, The Carribean Islands, Taiwan, the Phillipines, Malaysia, Indonesia, The Northern Territory of Australia and China. Additionally, the very virulent Foc Tropical Race 4 (FocR4) is now not limited to tropical regions in Asia but has alarmingly been diagnosed in Mozambique and Jordan and recently diagnosed in Northern Queensland indicating that the disease is emerging in new regions. Developing new strategies to overcome these emerging diseases can benefit from indepth knowledge and information on the interactive responses of its susceptible hosts.

Increasingly, transcriptome studies are being widely utilized to conduct investigation on gene expression in response to infection in banana. The breakthroughs of *Musa acuminata* and Musa balbisiana genomes have made ease in designing, collecting and analysing gene expression data. With this unlimited information the complexness of the disease infection network can be more accurately identified.

Based on reported studies most FOC pathogenesis studies have mainly focused on *Musa acuminata* cv Cavendish with little work reported on regional cultivars including cv Berangan. The objectives of this study will be to generate a large amount of RNA sequence data that would constitute a strong basis for transcriptomic studies in Musa accuminata cv Berangan by identifying the differentially expressed genes within the early stages of infection by FOC Race 4. For this purposes user will collect samples from systematically infected roots of soil grown plants under controlled conditions in the laboratory. Total RNA will be extracted from whole infected and non- infected roots at different time points early after challenge and a large number of short reads will be sequenced (RNA-Seq) to obtain qualitative and quantitative data of the transcriptome. This pre-field screenings assays has become an important tool as a key parameters that can be controlled and comparative gene expression analysis can be carried out to further understand the host response towards infection. This may also allow for early screening of breeding material for developing resistant varieties.

MATERIAL AND METHODS

Inoculum and Plant material

FocR4 (Isolate C1 HIR) will be isolated from the inner stem of an infected M. acuminata cv. 'Cavendish'.

Tissue culture-derived banana *Musa acuminata* cv. 'Berangan' plantlets will be obtained from local scientific laboratory. Plantlets were maintained in Murashige and Skoog (MS) medium for one month. For rooting, MS active charcoal (10g/L) will be used. Healthy plantlets aged 2 month with at least 3-5 green leaves, a minimum length of 5 cm of white rootsand stem diameters of between 0.5 to 1.0cm will be chosen for infection studies.

Sample collection

The infected plant root will be uprooted and washed with distilled water prior to deep freeze. Each biological replicates from different time points will be labelled separately and stored in -80°C.

RNA extraction and Quality control

Total RNA will be extracted with RNeasy® Plant Mini Kits, (Qiagen, Germany) according to the manufacturer`s instructions. DNA contamination will be eliminated using DNase. RNA quality will be evaluated using an Agilent 2100 Bioanalyzer with a minimum integrity number value of eight (8). mRNA purification kit will be used to purify mRNA with poly-A tail.

The purified RNA will be randomly fragmented and will reversed transcribed into cDNA. Adapters will be ligated into this fragment prior to PCR. 200- 400 bp fragment will be selected for paired end sequencing.

RNA Sequencing

Transcriptome analysis - Library preparation

The mRNA content will be recovered from total RNA using Illumina TruSeq RNA library Prep Kit. The mRNA will be captured twice on poly-T oligo magentic beads prior to fragmentation using fragmentation buffer. The fragmented strand will be used to synthesize the first cDNA strand by priming with random hexamers. The second strand will be generated and will be purified using Ampure XP beads (Illumina, UK). A single adenine base will be added to the 3` ends and sequencing adaptors will then ligated to the fragments and a flow cell will be used to select the range of fragments suitable for PCR amplification. The quality control analysis for the sample library and quantification of the DNA library template will be performed prior to sequencing. Sequencing will be carried out using an Illumina Hi Seq™ 2000 platform.

The quality control of the sequenced reads will be analyzed based on overal reads quality, total bases, total reads, GC content. Artificats such as low quality reads, adapter sequencde, contaminant DNA and PCR duplicates will preferably removed. Aligned reads will be e generated using TopHat prior alligned it against reference genome. Transcript assembly of aligned reads will be generated using cufflinks. Expressed profile will be calculated based on mapped transcript per sample. Normalization of transcript length and depth coverage will be carried out to compare expression profile between samples. Reads Per Kilobase of transcript per million mapped reads (RPKM) values will be used in normalization.

Processing mRNA sequence data

The raw RNA-Seg reads will be processed by FastQC (version: v0.10.0) to remove the low quality reads through a modular set of analyses and then mapped to PKWpseudochromosome genome (transcriptome re-seq reference genome in http://banana-genome.cirad.fr) using a fast splice junction mapper Tophat (version: v1.3.3). This aligns the RNA Seq reads to the reference genome (PKW_pseudochromosome) using the ultra high-throughput short read aligner Bowtieand then the mapping results will be analyzed to indentify the splice junctions between exons. Transcript abundance of the novel gene and expression level of mapped genes will be calculated with the program Cufflinks (version: v2.1.1). Gene expression levels will be normalized with reads per kilobase of exon per million mapped reads (RPKM) values. Aligned RNA sequence reads will be assembled into a parsimonious set of transcripts. Estimation of the relative abundance of these transcripts will be based on how many reads support each one by taking into account on the biases in library preparation protocols.

Bioinformatic Analysis

Computing resources, Programming skills, File formats, Quality Control

CLC Genomic Workbench software version will be used. NCBI database will be utilized (http://www.ncbi.nlm.nih.gov/genome/10976) to download genome sequences and gene annotation files. The reads will be functionally annotated using Blast2Go software. KEGG analysis will be carried out. Software such as Aspera, FileZilla and Blast2Go will be downloaded in LINUX system. Cross-platform file formats including fasta, fastq, sam, bam, gtf and gff files will commonly used. Raw data with adjunct sequences will be processed before being mapped. Basic tasks such as adapter removal, trimming quality will be set to 0.001 equivalent to Q30 and summary statistics on quality score will be performed by standard tools like Q30.

RNA Seq Analysis

Trimmed sequences of reads from RNA collected at 2, 48 and 96 hours from the infected samples will be analysed. The sequence will be mapped against the reference Musa genome sequence and also gene sequences. The experiment will be carried out through unpaired multi-group comparison using existing expression values (RPKM). Normalization will be carried out.

DGE analysis

The frequency of each DGE library will be statistically analysed to identify the differentially expressed genes. The false discovery rate (FDR) will be used to determine the threshold of P value in multiple tests by manipulating the FDR value. Gene expressions will be considered significantly different with a cut-off value FDR < 0.001 and $|\log2 \text{ ratio}| > 1$.

Venn diagram and Heat Map

Venn diagram will be generated using Venny based on pairwise comparison. The venn diagram will be drawn based on the read count. Heatmap will be generated using Rscript for heatmap.

Metabolic pathways

Gene Ontology and KEGG prediction

The transcript sequence will be extracted from banana gff files using per script. The extracted sequences will be search against the SwissProt database using BLASTX. Transdecoder will be used to predict the peptides from the transcript sequences to serve as an input into Trinotate SQLite DB. The BLASTX results will be loaded into Trinotate SQLite DB which will automatically linked to GO. Trinotate results will be generated with e-value threshold 0.000001. SwissProt ID will be extracted from

Trinotate table. The SwissProt ID will be mapped to KEGG Orthologous and KEGG pathway. All the results will be combined into the CLC GWB format using a custom scripts.

REFERENCES

1. 1stBASE Product Catalog (2009-2010).
2. Abbasi AR, Hajirezaei M, Hofius D, Sonnewald U, Voll LM (2007) Specific roles of α and γ tocopherol in abiotic stress responses of transgenic tobacco. Plant Physiol 143:1720-1738
3. Agrawal N, Dasaradhi, PVN, Mohammed A, Malhotra P, Bhatnagar RK, Mukherjee S (2003) RNA Interference:Biology, mechanisms and application. Microbiol Mol Biol Rev 67:657-685
4. Agriculture resource guide. Thermo Fisher Scientific, Applied Biosystem
5. AirClean Systems. The Fume Control Experts. Protecting the operator or process from chemical fumes, vapors and particulate.
6. Ajjawi I, Shintani D (2004) Engineered plants with elevated vitamin E: a nutraceutical success story. Trends Biotechnol 22:104-107
7. Andeani JK, Mohsenzadeh S, Mohabatkar H (2009) Isolation an characterization of partial DREB gene from four Iranian Triticum aestivum cultivars. World J Agric Sci 5:561-566

8. Article Shared by Parul Kumar
9. Asensi-Fabado MA, Munne-Bosch S (2010) Vitamins in plants: occurrence, biosynthesis and antioxidant. Trends in Plant Sci 15:582-592
10. Azzi A (2004) The role of α-tocopherol in preventing disease. Eur J Nutr 43:18-25
11. Bagci E, Kurt Aitzetmuller LB, Altan Y (2003) A chemotaxonomic approach to the fatty acid and tocochromanols content of Cannabis sativa L. (Cannabaceae). Turk J Bot 27:141-147
12. Bent A (2006) Arabidopsis thaliana floral dip transformation method. Methods Mol Biol 343: 87-103
13. Bermudez Y, Ahmadi S, Lowel NE, Kruk PA (2007) Vitamin E suppresses telomerase activity in ovarian cancer cells. Cancer Detection Prev 31:119-128
14. BGI (2015). Exome Sequencing Application Note
15. Brookman_Amissah N, Packer H, Prediger E, Sabel J (2012). QPCR Application Guide. Experimental Overview, Protocol, Troubeshooting. PrimeTime 3RD Edition.
16. Chadna R, Augustine R, Bisht NC (2012) Evaluation of candidate reference genes for gene expression normalization in Brassica juncea using real time quantitative RT-PCR. 7:1-10PLOS ONE
17. Cheng Z, Sattler S, Maeda H, Sakuragi Y, Bryant DA, DellaPenna D (2003) Highly divergent methyltransferase catalyze reaction in tocopherol and plastoquinone synthesis in cynobacteria and photosynthesis eukaryotes. Plant Cell 15:2343-2356
18. CI-110 Plant Canopy Imager. Bio-Science
19. CI-202 Portable Laser Leaf Area Meter. Bio-Science
20. CI-203 Handheld Laser Leaf Area Meter. Bio-Science
21. CI-340 Photosynthesis System. Bio-Science
22. CI-600 In-Situ Root Imager. Bio-Science
23. CI-710 Miniature Leaf Spectrophotometer. Bio-Science
24. Collakova E, DellaPenna D (2001) Isolation and functional analysis of homogentisate phytyltransferase from Synechocystis sp. PCC 6803 and Arabidopsis. Plant Physiol 127:1113-1124

25. Curtis IS (2004) Production of transgenic crops by the floral-dip method. Methods Mol Biol 286:103-109

26. Curtis IS, Nam H (2001) Transgenic radish (Raphanus sativus L. longipinnatus Bailey) by floral dip method plant development and surfactant are important in optimizing transformation efficiency. Transgenic Res 10:363-371

27. Dernburg AF, Zalevsky J, Colaiacovo MP (2000) Transgene-mediated co-suppression in the C. Elegans germ line. Genes Dev 14:1578-1583

28. Dudareva N, Pichersky E (2008) Metabolic engineering of plant volatiles. Curr Opin Biotechnol 19:1-9

29. Durai M, Dubey SC, Tripathi A, 2012. Genetic diversity analysis and development of SCAR marker for detection of Indian populations of Fusarium oxysporum f. sp. ciceris causing chickpea wilt Folia Microbiology 57: 229–235.

30. Dutta A, Singh M (2011) Comparative analysis of aqueous extracts of amaranth and coriander in scavenging free radical activity and protection of DNA against oxidative damage. Chiang Mai J Sci 38:560-571

31. Endrigkeit J, Wang X, Cai D, Zhang C, Long Y, Meng J, Jung C (2009) Genetic mapping, cloning and functional characterization of the BnaX.VTE4 gene encoding a γ-tocopherol metyltransferase from oilseed rape. Theor Appl Genet 119:567-575

32. Engel KH, Frenzel TH, Miller A (2002) Current and future benefits from the use of GM technology in food production. Toxicol Lett 127:329-336

33. Eppendorf (2016). Liquid Handling, Sample Handling, Cell Handling Catalog 2016

34. Esco World Class Worldwide.

35. Falk F, Munne-Bosch S (2010) Tocochromanol functions in plants: antioxidation and beyond. J Exp Bot 61:1549-1566

36. Fukusaki E, Kobayashi A (2005) Plant metabolomics: Potential for practical operation. J Biosci Bioeng 100:347-354

37. Gasior R, Pieszka M, Brzoska F (2009) Validation of a method dor simultaneous determination of tocopherols and tocotrienols in

cereals using normal phase HPLC. J Anim Feed Sci 18:173-192

38. Gasteiger E, Hoogland C, Gattiker A, Duvaud S, Wilkins M.R, Appel R.D, Bairoch A (2005) *Protein Identification and Analysis Tools on the ExPASy Server* (In) John M. Walker (ed): The Proteomics Protocols Handbook, Humana Press pp. 571-607
39. GE Healthcare Life Sciences (2013). In Cell Analyzer High-Content Cellular Analysis System.
40. Grabowska A, Filipecki M (2004) Infiltration with Agrobacterium the method for stable transformation avoiding tissue culture. Acta Physiol Plant 26:451-458
41. HANNA instruments. Agriculture & Aquaculture.
42. Harrison SJ, Mott EK, Parsley K, Aspinall S, Gray JC, Cottage A (2006) A rapid and robust method of identifying transformed Arabidopsis thaliana seedling following floral dip transformation. Plant Methods 2:1-7
43. Heathrow Scientific (2012). Design, Function, Innovation.
44. Hellens RP, Allan AC, Friel EN, Bolitho K, Grafton K, Templeton MD, Karunairethnam, Gleave AP, Liang WA (2005) Transient expression for functional genomics, quantification of promoter activity and RNA silencing in plants. Plant Methods 1:1-14
45. Herbers K (2003) Vitamin production in transgenic plants. J Plant Physiol 160:821-829
46. HiMedia (2008). Plant Tissue Culture Manual.
47. Hofius D, Hajirezaei MR, Geiger M, Tschiersch H, Melzer M, Sonnewald U (2004) RNAi-mediated tocopherol deficiency impairs photoassimilate export in transgenic potato plants. Plant Physiol 135:1256-1268
48. Horvath G, Wessjohann L, Bigirimana J, Monica H, Jansen M, Guisez Y, Caubergs R, Horemans N (2006) Accumulation of tocopherols and tocotrienols during seed development of grape (Vitis vinifera L. cv Albert Lavalle). Plant Physiol Biochem 44:724-731
49. Hunter SC, Cahoon EB (2007) Enhancing vitamin E in oilseeds: Unravelling tocopherol and tocotrienol biosynthesis. Lipids 42:97-108

50. Illumina (2014). Next Generation Sequencing Application Guide.
51. Islam MS and Shuva NZ (2010). Barcode technology and its use and applications: Astudy of selected libraries of Bangladesh. The international Information & Library Review 42:27-33
52. Jian B, Hou W, Wu C, Liu B, Liu W, Song S, Bi Y, Han T (2009) Agrobacterium rhizogenes mediated transformation of Superrot-derived Lotus corniculus plants: a valuable tool for functional genomics. BMC Plant Biol 9:1-14
53. Jiang Q, Freiser H, Wood KV, Yin X (2007) Identification and quantification of novel vitamin E metabolites, sulfated long-chain carboxychromanols, in human A549 cells and in rats. Lipids Res 48:1221-1230
54. Jimenez-Fernandez, D., M. Montes-Borrego, J.A. Navas-Cortes, R.M. Jimenez-Diaz, B.B. Landa. 2010. Identification and quantification of Fusarium oxysporum in planta and soil by means of an improved specific and quantitative PCR assay. Appl. Soil Eco. 46:372-382.
55. Johnson M, Zaretskaya I, Raytselis Y, Merezhuk Y, McGinnis S, Madden TL (2008) NCBI BLAST: a better web interface. Nucleic Acids Res 36:5-9
56. Juan G, Liu G, Shuang Aly AA (2009). Vitamin E metabolic modulation in plants IN Herbal Drugs: Ethnomedicine to Modern Medicine, Springer-Verlag Berlin Heidelberg pp. 333-352
57. Ko MK, Soh H, Kim KM, Kim YS (2007) Stable production of transgenic pepper plants mediated by Agrobacterium tumefaciens. HortSci 42:1425-1430
58. Korchazhkina O, Jones E, Czauderna M, Spencer SA, Kowalczyk J (2006) HPLC with UV detection for measurement of vitamin E in human milk. Acta Chromatogr 16:48-57
59. Kruk J, Szymanska R, Krupinska K (2008) Tocopherol quinone content of green algae and higher plants revised by a new high-sensitive fluorescence detection method using HPLC-effects of high light stress and senescence. J. Plant Physiol 165:1238-1247
60. Kumar KK, Maruthasalam S, Loganathan M, Sudhakar D, Balasubramaniam P (2005) An improved Agrobacterium-mediated transformation protocol for recalcitrant elite Indica rice cultivars.

Plant Mol Biol Report 23:67-73

61. Kuta DD, Tripathi L (2005) Agrobacterium induced hypersentive necrotic reaction in plant cells: a resistance response against Agrobacterium mediated DNA transfer. Afr J Biotechnol 4:752-757
62. Laskowski RA, Watson JD, Thornton JM (2005) ProFunc: a server for predicting protein function from 3D structure. Nucleic Acids Res 33:89-93
63. Lee K, Lee SM, Park SR, Jung J, Moon JK, Cheong JJ, Kim M (2007) Overexpression of Arabidopsis homogentisate phytyltransferase ot tocopherol cyclase elevates vitamin E content by increasing gamma tocopherol level in lettuce (Lactuca sativa L.). Mol Cells 24:301-306
64. Lee MW, Yang Y (2006) Transient expression assay by agroinfiltration of leaves. Methods Mol Biol 9: 323-225
65. Li C, Shao J, Wang Y, Li W, Guo D, Yab B, Xia Y, Peng M (2013). Analysis of banana transcriptome and global gene expression profiles in banana roots in response to infection by race 1 and tropical race 4 of Fusarium oxysporum f. sp. cubense. BMC Genomics, 14: 851-867.
66. Lin J, Zhou B, Yang Y, Mei J, Zhao X, Guo X, Huang X, Tang D, Liu X (2009) Piercing and vacuum infiltration of the mature embryo: a simplified method for Agrobacterium mediated transformation of Indica rice. Plant Cell Rep 28:1065-1074
67. Lin YH., Su CC., Chao CP., Chen CY., Chang CJ., Huang JW., Linda Chang PF (2013) Amolecular diagnosis method using real time PCR for quantification and detection of Fusarium oxysporum f. sp. cubense race 4. Eur J Plant Pathol 135:395-405.
68. Ling MT, Luk SU, Al-Ejeh F, Khanna KK (2012) Tocotrienol as a potential anticancer agent. Carcinog 33:233-239
69. Lodge JK, Traber MG, Elsner A, Brigelius-Flohe R (2000) A rapid method for the extraction and determination of vitamin E metabolites in human urine. J Lipid Res 41:148-154
70. Lopez-Mondejar et al., 2011
71. Louime C, Vasanthaiah HKN, Jittayasothorn Y, Lu J, Basha SM, Thipyapong P, Boonkerd N (2008) A simple and efficient protocol

for high quality RNA extraction and cloning of Chalcone synthase partial cds from Muscadine grape cultivars (Vitis Rotundifolia Michx.). Eur J Sci Res 22:232-240

72. Martinez-Trujillo M, Limones-Briones V, Cabera-Ponce JL, Herrera-Estrella (2004) Improving transformation efficiency of Arabidopsis thaliana by modifying the floral dip method. Plant Mol Biol Report 22:63-70
73. Meter Toledo. Instrument in motion-powered by Seven
74. Mishra RK, Pandey BK, Muthukumar M, Pathak N, Zeeshan m (2013) Detection of Fusarium wilt pathogens of Psidium guajava L. In soil using culture independent PCR (ciPCR) Saudi J of biological Sciences 20:51-56
75. Mukherjee S, Zhang Y (2011) Protein-protein complex structure prediction by multimeric threading and template recombination. Struct 19:955-966
76. New England Biolabs. Genome Editing. Tools to support CRISPR/ CASP Applications
77. Oka SI, Midorikawa K, Kodama H (2010) Co-suppression and RNAi induced by Arabidopsis ortholog gene sequences in tobacco. Plant Biotechnol Rep 4:185-192
78. Olimpieri I, Caccia R, Picarella ME, Pucci A, Santangelo E, Soressi GP (2011) Constitutive co-suppression of the *GA 20-oxidase1* gene in tomato leads to severe defects in vegetative and reproductive development. Plant Sci180:496-503
79. Oxford (2008) Dictionary of Biology. Oxford University Press Inc, New York Petersen et al., 2011
80. Pieper U, Webb BM, Barkan DT, Schneidman-Duhovny D, Schlessinger A, Braberg H, Yang Z, Meng EC, Pettersen EF, Huang CC, Datta RS, Sampathkumar P, Madhusudhan MS, Sjolander K, Ferrin TE, Burley SK, Sali A (2011) ModBase, a database of annotated comparative protein structure modelsandand associated resources. Nucleic Acids Res 39:465-474
81. Potenza C, Aleman L, Sengupta-Gopalan C (2004) Targeting transgene expression in research,
82. PrimePCR Assays AND Panels FOR Real-Time PCR

83. Raclaru M, Gruber J, Kumar R, Sadre R, Luhs W, Karim Zarhloul M, Friedt W, Frentzen M, Weier D (2006) Increase of the tocochromanol content in transgenic Brassica napus seeds by overexpression of key enzymes involved in prenylquinone biosynthesis. Mol Breed 18:93-107

84. Raja K (2011) Bio-programming prospects of JAVA: a computttional move towards the understanding of the biological aspects of genes and proteins. Indian J Computer Sci Eng 2:949-956

85. Rajesh S, Manickam A (2006) Prediction of functions for two LEA proteins from mung bean. Bioinforma 1:133-138

86. *Rost B Yachdav G Liu J (2004) The PredictProtein Server. Nucleic Acids Research 32(Web Server issue):W321-W326*

87. Ruperez FJ, Martin D, Herrera E, Barbas C (2001) Chromatographic analysis α-tocopherol and related compounds in various matrices. J Chromatogr A 935: 45-69

88. Scarlett K, Tesoriero L, Daniel R, Guest D (2013) Detection and quantification of Fusarium oxysporum F. sp. cucumerinum in environmental samples using a specific quantitative PCR assay. Eur Plant Pathol 137:315-324

89. Semchuk NM, Lushchak OV, Falk J, Krupinska K, Lushchak VI (2009) Inactivation of genes, encoding tocopherol biosynthetic pathway enzymes, results in oxidative stress I outdoor grown Arabidopsis thaliana. Plant Physiol Biochem 384-390

90. Sen CK, Khanna S, Roy S (2007) Tocotrienols in health and disease: The other half of the natural vitamin E family. Mol Asp Med 28:692-728

91. Seo YS, Kim SJ, Harn CH, Kim WT (2011) Ecotopic expression of apple fruit homogentisate phytyltransferase gene (MdHPT1) increases tocopherol in transgenic tomato (Solanum lycopersicum cv. Micro-Tom) leaves and fruits. Phytochem 72:321-329

92. Singh T, Biswas D, Jayaram B (2011) AADS-An automated active site identification, dockingandand scoring protocol for protein targets based on physicochemical descriptors. J Chem Inf Model 5:2515-2527

93. Sivakumar G, Bacchetta L, Gatti R, Zappa G (2005) HPLC screening of natural vitamin E from Mediterranean plant biofactories-a basic

tool for pilot-scale bioreactors production of α-tocopherol. J Plant Physiol 162:1280-1283

94. Storozhenko S, Ravanel S, Zhang GF, Rebeille F, Lambert W, Van Der Straten D (2005) Folate enhancement in staple crops by metabolic engineering. Trends Food Sci Technol 16:271-281

95. Stothard P (2000) The Sequence Manipulation Suite: JavaScript programs for analyzing and formatting protein and DNA sequences. Biotech 28:1102-1104

96. Sundram K, Sambanthamurthi R, Tan YA (2003) Palm fruit chemistry and nutrition. Asia Pacific J Clin Nutr 12:355-362

97. Tague BW, Mantis J (2006) In planta Agrobacterium mediated transformation by vacuum infiltration. Methods Mol Biol, vol 323: Arabidopsis Protocols, second Edition. Humana Press Inc., Totowa NJ

98. Tahrin Mahmood and Ping-Chang Yang Western Blot: Technique, Theoryandand Trouble Shooting by North American Journal of Medical Sciences

99. Tamura K, Peterson D, Peterson N, Stecher G, Nei M, Kumar S (2011) MEGA5: Molecular evolutionary genetics analysis using maximum likelihood, evolutionary distance and maximum parsimony. Methods Mol Biol Evol Article in Press

100. Tavva VS, Kim YH, Kagan IA, Dinkins RD, Kim KH, Collins GB (2007) Increased α-tocopherol content in soybean seed overexpressing the Perilla frutescens γ-tocopherol methyltransferase. Plant Cell Rep 26:61-70

101. The LAB Shop. Interscience SDN BHD

102. Trucco F,Tranel J (2011) Amaranthus. In Wild Crop Relatives: Genomics and Breeding Resources, Vegetables. Springer-Verlag Berlin Heidelberg. C. Kole (ed.)

103. Ufaz S, Galili G (2008) Improving the content of essential amino acids in crop plants: Goals and opportunities. Plant Physiol 147:954-961

104. Umaiyal Munusamy, Siti Nor Akmar Abdullah, Maheran Abd Aziz, Huzwah Khaza`ai. (2015). Metabolic engineering of α-tocotrienol through PTGS mechanisms and isoprenoid/non-mevalonate

pathways in perennial crops. Plant Cell Biotechnology and Molecular Biology, 16:119-129

105. Umaiyal Munusamy, Siti Nor Akmar Abdullah, Maheran Abd Aziz, Huzwah Khaza`ai. And Lai Mun Seong (2013). Induced production of α-tocotrienol by co-suppression of tocopherol cyclase gene in vegetable crops. American Journal of Biochemistry and Biotechnology 9: 355-364
106. Umaiyal Munusamy, Siti Nor Akmar Abdullah, Maheran Abd Aziz, Huzwah Khaza`ai. (2013). Female reproductive system of Amaranthus as the target for Agrobacterium-mediated transformation. Advances in Bioscience and Biotechnology 4:188-192
107. Valentin HE, Qi Q (2005). Biotechnological production and application of vitamin E: current state and prospects. Appl Microbiol Biotechnol 68:436-444
108. Van Eenennaam AL, Li G, Venkatramesh M, Levering C, Gong X, Jamieson AC, Rebar EJ, Shewmaker CK, Case CC (2004) Elevation of seed α-tocopherol levels using plant-based transcription factors targeted to an endogenous locus. Metabolic Eng, 6:101-108
109. Viola V, Pilolli F, Piroddi M, Pierpaoli E, Orlando F, Provincialli M, Betti M, Mazzini F, Galli F (2011b) Why tocotrienols work better: insights into the in vitro anti cancer mechanism of vitamin E. Genes Nutrition vitamin E. Tocopherols from various sources. Biol Chem 122,99-107
110. Voll LM, Abbasi AL (2007) Are there specific In vivo roles for α- and γ-tocopherol in plants?. Plant Signal Behav 2:486-488Weckwerth and Fiehn, 2002)
111. Welch RM (2002) The impact of mineral nutrients in food crops on global human health. Plant Soil 247:83-90
112. Wise AA, Liu Z, Bins AN (2006) Three methods for the introduction of foreign DNA into Agrobacterium. Methods Mol Biol 343: 43-53
113. Xu R, Li QQ (2008) Protocol: Streamline cloning of genes into binary vectors in Agrobacterium via the Gateway®TOPO vector system. Plant Methods 4:2-7
114. Xu Z (2002) Analysis of tocopherol. Curr Protoc Food Anal Chem D1.5.1-D1.5.12

115. Yang A, Su Q, An L (2009) Ovary-drip transformation: a simple method for directly generating vector and marker free tranagenic maize (Zea mays L.) with linear GFP cassette transformation. Planta 229:793-801

116. Yasmeen A, Mirza B, Inayatullah S, Safdar N, Jamil M, Ali S, Fayyaz Choudhry M (2009) In planta transformation of tomato 27:20-

117. You CS, Sontag TJ, Swanson JE, Parker RS (2005) Long-chain carboxychromanols are the major metabolites of tocopherols and tocotrienols in A549 lung epithelial cells but not HepG2 cells. Nutr Metab pp: 227-232

118. Zale JM, Agarwal S, Loar S, Steber CM (2009) Evidence for stable transformation of wheat by floral dip in Agrobacterium tumefaciens. Plant Cell Rep 28:903-913

119. Zhang X, Henriques R, Lin SS, Niu Q, Chua NH (2006) Agrobacterium mediated transformation of Arabidopsis thaliana using the floral dip method. Nature Protoc 1:1-6

INDEX